T0246407

Animal Genetics and Breeding

Edited by
Zaiden Jonston

Larsen & Keller
www.larsen-keller.com

Animal Genetics and Breeding
Edited by Zaiden Jonston
ISBN: 978-1-63549-026-8 (Hardback)

Published by Larsen and Keller Education,
5 Penn Plaza,
19th Floor,
New York, NY 10001, USA

Cataloging-in-Publication Data

Animal genetics and breeding / edited by Zaiden Jonston.
p. cm.
Includes bibliographical references and index.
ISBN 978-1-63549-026-8
1. Animal genetics. 2. Animal breeding. 3. Artificial insemination. I. Jonston, Zaiden.
QH432 .A55 2017
591.35--dc23

Printed and bound in the United States of America.

For more information regarding Larsen and Keller Education and its products, please visit the publisher's website www.larsen-keller.com

Table of Contents

Preface

This book is a valuable compilation of topics, ranging from the basic to the most complex theories and principles in the field of animal genetics and breeding. It explores all the important aspects of animal genetics and breeding in the present day scenario. Animal genetics studies selected breeding of livestock through genetic intervention with the purpose of increasing genetic value. The book also talks about the basic principles of breeding, giving details about animal breeding, husbandry and breeding in the wild. Such selected concepts that redefine this field have been presented in this text. For all those who are interested in animal genetics and breeding, this textbook can prove to be an essential guide.

A short introduction to every chapter is written below to provide an overview of the content of the book:

Chapter 1 - Genetics is the study of genes and of inheritance. Genetics is an emerging field of study; the following chapter will not only provide an overview, it will also delve deep into the variegated topics related to it. This chapter is an introductory chapter, which will introduce briefly all the significant aspects of animal breeding and genetics; **Chapter 2 -** The chapter closely examines the key concepts of genetics to provide an extensive understanding of the subject. Some of the key concepts discussed in this chapter are chromosomes, genetic diversity, mutation, DNA and RNA. A chromosome is a structure that contains the DNA of a living organism whereas genetic diversity is the sum of all the genetic characteristics in the genetic nature of an organism. The text strategically encompasses and incorporates the major key concepts of genetics, providing a complete understanding; **Chapter 3 -** Breeding is the function of reproduction and is concerned with particularly animals and plants. Some of the aspects of breeding discussed are captive breeding, seasonal breeder, breeding pair, inbreeding, outcrossing, backcrossing and culling. This section is an overview of the subject matter incorporating all the major aspects of breeding; **Chapter 4 -** A gene is the functional unit of heredity. It is very important to understand gene nomenclature, gene expression, essential gene and heredity, to have a profound understanding of genes. This chapter has been carefully written to provide an easy understanding of the varied facets of genes; **Chapter 5 -** The process of reproducing by means other than sexual intercourse is referred to as artificial insemination. It's a deliberate introduction of sperm into the female's uterus. As compared to the natural way of reproducing, artificial insemination is expensive and involves professional help. The application of artificial insemination in the current scenario has been thoroughly discussed in this chapter.

Finally, I would like to thank my fellow scholars who gave constructive feedback and my family members who supported me at every step.

Editor

1

Introduction to Animal Breeding and Genetics

Genetics is the study of genes and of inheritance. Genetics is an emerging field of study; the following chapter will not only provide an overview, it will also delve deep into the variegated topics related to it. This chapter is an introductory chapter, which will introduce briefly all the significant aspects of animal breeding and genetics.

Animal Breeding

Animal breeding is a branch of animal science that addresses the evaluation (using best linear unbiased prediction and other methods) of the genetic value (estimated breeding value, EBV) of livestock. Selecting for breeding animals with superior EBV in growth rate, egg, meat, milk, or wool production, or with other desirable traits has revolutionized livestock production throughout the world. The scientific theory of animal breeding incorporates population genetics, quantitative genetics, statistics, and recently molecular genomics and is based on the pioneering work of Sewall Wright, Jay Lush, and Charles Henderson.

Breeding Stock

Breeding stock is a group of animals used for the purpose of planned breeding. When individuals are looking to breed animals, they look for certain valuable traits in purebred animals, or may intend to use some type of crossbreeding to produce a new type of stock with different, and presumably superior abilities in a given area of endeavor. For example, when breeding swine the "breeding stock should be sound, fast growing, muscular, lean, and reproductively efficient." The "subjective selection of breeding stock" in horses has led to many horse breeds with particular performance traits.

Purebred Breeding

Mating animals of the same breed for maintaining such breed is referred to as purebred breeding. Opposite to the practice of mating animals of different breeds, purebred breeding aims to establish and maintain stable traits, that animals will pass to the next generation. By "breeding the best to the best," employing a certain degree of inbreeding, considerable culling, and selection for "superior" qualities, one could develop a bloodline or "breed" superior in certain respects to the original base stock.

Such animals can be recorded with a breed registry, the organisation that maintains pedigrees and/or stud books. The observable phenomenon of hybrid vigor stands in contrast to the notion of breed purity.

Backyard Breeding

In the United States, a backyard breeder is someone who breeds animals, often without registration and with a focus on profit. In some cases the animals are inbred narrowly for looks with little regard to health. The term is considered derogatory. If a backyard dog breeder has a significant number of breeding animals, they become associated with puppy mills. Most puppy mills are licensed with the USDA.

Purebred

Purebreds, also called purebreeds, are cultivated varieties or *cultivars* of an animal species, achieved through the process of selective breeding. When the lineage of a purebred animal is recorded, that animal is said to be *pedigreed.*

The term *purebred* is occasionally confused with the proper noun *Thoroughbred,* which refers exclusively to a specific breed of horse, one of the first breeds for which a written national stud book was created since the 18th century. Thus a purebred animal should never be called a "thoroughbred" unless the animal actually is a registered Thoroughbred horse.

True Breeding

In the world of selective animal breeding, to "breed true" means that specimens of an animal breed will breed true-to-type when mated like-to-like; that is, that the progeny of any two individuals in the same breed will show consistent, replicable and predictable characteristics. A puppy from two purebred dogs of the same breed, for example, will exhibit the traits of its parents, and not the traits of all breeds in the subject breed's ancestry.

However, breeding from too small a gene pool, especially direct inbreeding, can lead to the passing on of undesirable characteristics or even a collapse of a breed population due to inbreeding depression. Therefore, there is a question, and often heated controversy, as to when or if a breed may need to allow "outside" stock in for the purpose of improving the overall health and vigor of the breed.

Because pure-breeding creates a limited gene pool, purebred animal breeds are also susceptible to a wide range of congenital health problems. This problem is especially prevalent in competitive dog breeding and dog show circles due to the singular emphasis on aesthetics rather than health or function. Such problems also occur within certain segments of the horse industry for similar reasons. The problem is further compounded when breeders practice inbreeding. The opposite effect to that of the restricted gene pool caused by pure-breeding is known as hybrid vigor, which generally results in healthier animals.

Pedigrees

A pedigreed animal is one that has its ancestry recorded. Often this is tracked by a major registry. The number of generations required varies from breed to breed, but all pedigreed animals have papers from the registering body that attest to their ancestry.

The word "pedigree" appeared in the English language in 1410 as "pee de Grewe", "pedegrewe" or "pedegru", each of those words being borrowed to the Middle French "pié de grue", meaning "crane foot". This comes from a visual analogy between the trace of the bird's foot and the three lines used in the English official registers to show the ramifications of a genealogical tree.

A purebred Arabian horse.

Sometimes the word *purebred* is used synonymously with *pedigreed*, but purebred refers to the animal having a known ancestry, and pedigree refers to the written record of breeding. Not all purebred animals have their lineage in written form. For example, until the 20th century, the Bedouin people of the Arabian peninsula only recorded the ancestry of their Arabian horses via an oral tradition, supported by the swearing of religiously based oaths as to the asil or "pure" breeding of the animal. Conversely, some animals may have a recorded pedigree or even a registry, but not be considered "purebred". Today the modern Anglo-Arabian horse, a cross of Thoroughbred and Arabian bloodlines, is considered such a case.

Purebreds by Animal

Purebred Dogs

Pure-breed Zwergschnauzer

A *purebred dog* is a dog of a modern breed of dog, with written documentation showing the individual purebred dog's descent from its breeds' foundation stock. In dogs, the term breed is used two ways: loosely, to refer to dog types or landraces of dog (also called natural breeds or ancient breeds); or more precisely, to refer to modern breeds of dog, which are documented so as to be known to be descended from specific ancestors, that closely resemble others of their breed in appearance, movement, way of working and other characters; and that reproduce with offspring closely resembling each other and their parents. Purebred dogs are breeds in the second sense.

New breeds of dog are constantly being created, and there are many websites for new breed associations and breed clubs offering legitimate registrations for new or rare breeds. When dogs of a new breed are "visibly similar in most characteristics" and have reliable documented descent from a "known and designated foundation stock", then they can then be considered members of a breed, and, if an individual dog is documented and registered, it can be called *purebred.*

Purebred Horses

The domestication of the horse resulted in a small number of domesticated stallions (possibly a single male ancestor) being crossed on wild mares that had adapted to local conditions. This ultimately produced horses of four basic body types, once thought to be wild prototypes, but now considered to be landraces. Many of these animals were then bred true to original type by selected breeding, though emphasizing certain inherent traits (such as a good temperament, suitable to training by humans) to a greater degree than others. In other cases, horses of different body types were cross bred until a desired characteristic was achieved and bred true.

Written and oral histories of various animals or pedigrees of certain types of horse have been kept throughout history, though breed registry stud books trace only to about the 13th century, at least in Europe, when pedigrees were tracked in writing, and the practice of declaring a type of horse to be a breed or a purebred became more widespread.

Certain horse breeds, such as the Andalusian horse and the Arabian horse, are claimed by aficionados of the respective breeds to be ancient, near-pure descendants from an ancient wild prototype, though mapping of the horse genome as well as the mtDNA and y-DNA of various breeds has largely disproved such claims.

Purebred Cats

A cat whose ancestry is formally registered is called a pedigreed or purebred cat. Technically, a *purebred* cat is one whose ancestry contains only individuals of the same breed. A *pedigreed* cat is one whose ancestry is recorded, but may have ancestors of different breeds.

Purebred Himalayan

The list of cat breeds is quite large: most cat registries actually recognize between 30 and 40 breeds of cats, and several more are in development, with one or more new breeds being recognized each year on average, having distinct features and heritage. Owners and breeders compete in cat shows to see whose animal bears the closest resemblance (best *conformance*) to an idealized definition, based on breed type and the breed standard for each breed.

Because of common crossbreeding in populated areas, many cats are simply identified as belonging to the mixed types of domestic long-haired and domestic short-haired cat, depending on their type of fur.

Some original cat breeds that have a distinct phenotype that is the main type occurring naturally as the dominant domesticated cat type in their region of origin are sometimes considered subspecies and in the past received names as such, although this is no longer supported by feline biologists. with *F. silvestris* the wild cat and *F. s. catus* the domestic. Some of these cat breeds (with their invalid scientific names for historical interest) are:

- *F. catus anura* - the Manx
- *F. catus siamensis* - the Siamese
- *F. catus cartusenensis* - the Chartreux
- *F. catus angorensis* - the Turkish Angora

Purebred Livestock

Most domesticated farm animals also have true-breeding breeds and breed registries, particularly cattle, sheep, goats, rabbits, and pigs. While animals bred strictly for market sale are not always purebreds, or if purebred may not be registered, most livestock producers value the presence of purebred genetic stock for the consistency of traits such animals provide. It is common for a farm's male breeding stock in particular to be of purebred, pedigreed lines.

Purebred Barbados Blackbelly hair sheep.

In cattle, some breeders associations make a difference between "purebred" and "full blood". Full blood cattle are fully pedigreed animals, where every ancestor is registered in the herdbook and shows the typical characteristics of the breed. Purebred are those animals that have been bred-up to purebred status as a result of using full blood animals to cross with an animal of another breed. The breeders association rules the percentage of fullblood genetics required for an animal to be considered purebred, usually above 87.5%.

Charolais calves that were transferred as embryos, with their Angus and Hereford recipient mothers.

Artificial breeding via artificial insemination or embryo transfer is often used in sheep and cattle breeding to quickly expand, or improve purebred herds. Embryo transfer techniques allow top quality female livestock to have a greater influence on the genetic advancement of a herd or flock in much the same way that artificial insemination has allowed greater use of superior sires.

Wild Species, Landraces, and Purebred Species

Breeders of purebred domesticated species discourage crossbreeding with wild species, unless a deliberate decision is made to incorporate a trait of a wild ancestor back into a given breed or strain. Wild populations of animals and plants have evolved naturally over millions of years through a process of natural selection in contrast to human controlled Selective breeding or artificial selection for desirable traits from the human point of view. Normally, these two methods of reproduction operate independently of one another. However, an intermediate form of selective breeding, wherein animals or plants are bred by humans, but with an eye to adaptation to natural region-specific conditions and an acceptance of natural selection to weed out undesirable traits, created many ancient domesticated breeds or types now known as landraces.

Many times, domesticated species live in or near areas which also still hold naturally evolved, region-specific wild ancestor species and subspecies. In some cases, a domesticated species of plant or animal may become feral, living wild. Other times, a wild species will come into an area inhabited by a domesticated species. Some of these situations lead to the creation of hybridized plants or animals, a cross between the native species and a domesticated one. This type of crossbreeding, termed genetic pollution by those who are concerned about preserving the genetic base of the wild species, has become a major concern. Hybridization is also a concern to the breeders of purebred species as well, particularly if the gene pool is small and if such crossbreeding or hybridization threatens the genetic base of the domesticated purebred population.

The concern with genetic pollution of a wild population is that hybridized animals and plants may not be as genetically strong as naturally evolved region specific wild ancestors wildlife which can survive without human husbandry and have high immunity to natural diseases. The concern of purebred breeders with wildlife hybridizing a domesticated species is that it can coarsen or degrade the specific qualities of a breed developed for a specific purpose, sometimes over many generations. Thus, both purebred breeders and wildlife biologists share a common interest in preventing accidental hybridization.

Crossbreed

A crossbreed or crossbred usually refers to an organism with purebred parents of two different breeds, varieties, or populations. *Crossbreeding*, sometimes called "designer crossbreeding", refers to the process of breeding such an organism, often with the intention to create offspring that share the traits of both parent lineages, or producing an organism with hybrid vigor. While crossbreeding is used to maintain health and viability of organisms, irresponsible crossbreeding can also produce organisms of inferior quality or dilute a purebred gene pool to the point of extinction of a given breed of organism.

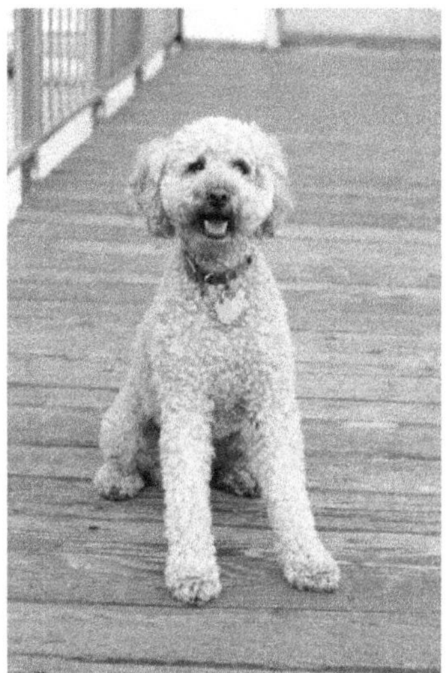

A Labradoodle, a crossbreed between a poodle and a retriever

The term is also used at times to refer to a domestic animal of unknown ancestry where the breed status of only one parent or grandparent is known, though the term "mixed breed" is technically more accurate. The term outcross is used to describe a type of crossbreeding used within a purebred breed to increase the genetic diversity within the breed, particularly when there is a need to avoid inbreeding.

In animal breeding, *crossbreed* describes crosses within a single species, while *hybrid* refers to crosses between different species. In plant breeding terminology, the term *crossbreed* is uncommon. No universal term is used to distinguish hybridization or crossing within a population from those between populations, or even those between species.

Crossbreeds in Specific Animals

Cats

The many newly developed and recognized breeds of domestic cat are crossbreeds between existing, well-established breeds (sometimes with limited hybridization with some wild species), to either combine selected traits from the foundation stock, or propagate a rare mutation without excessive inbreeding. However, some nascent breeds such as the Aegean cat are developed entirely from a local landrace population. Most experimental cat breeds are crossbreeds.

Cattle

In cattle, there are systems of crossbreeding. In many crossbreeds, one is larger than the other. One is used when the purebred females are particularly adapted to a specific environment, and are crossed with purebred bulls from another environment to produce a generation having traits of both parents.

Sheep

The large number of breeds of sheep, which vary greatly, creates an opportunity for crossbreeding to be used to tailor production of lambs to the goal of the individual stockman.

Llamas

Results of crossbreeding classic and woolly breeds of llama are unpredictable. The resulting offspring displays physical characteristics of either parent, or a mix of characteristics from both, periodically producing a fleeced llama. The results are increasingly unpredictable when both parents are crossbreeds, with possibility of the offspring displaying characteristics of a grandparent, not obvious in either parent.

Dogs

A *crossbred* dog is a cross between two (sometimes more) known breeds, and is usually distinguished from a *mixed-breed dog*, which has ancestry from many sources, some of which may not be known. Crossbreeds are popular, due to the belief that they have increased vigor without loss of attractiveness of the dog. Certain planned crossbreeding between purebred dogs of different breeds can produce puppies worth more than their purebred parents, due to a high demand.

Horses

Crossbreeding in horses is often done with the intent of ultimately creating a new breed of horse. One type of modern crossbreeding in horses is used to create many of the warmblood breeds. Warmbloods are a type of horse used in the sport horse disciplines, usually registered in an open stud book by a studbook selection procedure that evaluates conformation, pedigree and, in some animals, a training or performance standard. Most warmblood breeds began as a cross of draft horse breeds on Thoroughbreds, but have, in some cases, developed over the past century to the point where they are considered to be a true-breeding population and have a closed stud book. Other types of recognized crossbreeding include that within the American Quarter Horse, which will register horses with one Thoroughbred parent and one registered Quarter Horse parent in the "Appendix" registry, and allow such animals full breed registration status as Quarter Horses if they meet a certain performance standard. Another well-known crossbred horse is the Anglo-Arabian, which may be produced by a purebred Arabian horse crossed on a Thoroughbred, or by various crosses of Anglo-Arabians with other Anglo-Arabians, as long as the ensuing animal never has more than 75% or less than 25% of each breed represented in its pedigree.

The National Show Horse was developed from crossbreeding programs in the 1970s and 1980s that blended Arabian horse and American Saddlebred bloodlines

Hybrid Animals

A hybrid animal is one with parentage of two separate species, differentiating it from crossbred animals, which have parentage of the same species. Hybrids are usually, but not always, sterile.

One of the most ancient types of hybrid animal is the mule, a cross between a female horse and a male donkey. The liger is a hybrid cross between a male lion and female tiger. The yattle is a cross between a cow and a yak. Other crosses include the tigon (between a male tiger and female lion) and yakalo (between a yak and buffalo). The Incas recognized that hybrids of Lama glama (llama) and Lama pacos (alpaca) resulted in a hybrid with none of the advantages of either parent.

At one time it was thought that dogs and wolves were separate species, and the crosses between dogs and wolves were called wolf hybrids. Today wolves and dogs are both recognized as *Canis lupus*, but the old term "wolf hybrid" is still used.

Mixed Breeds

A mixed-breed animal is defined as having undocumented or unknown parentage, while a crossbreed generally has known, usually purebred parents of two distinct breeds or varieties. A dog of unknown parentage is often called a mixed-breed dog, "mutt" or "mongrel." A cat of unknown parentage is often referred to as domestic short-haired or domestic long-haired cat generically, and in some dialects is often called a "moggy". A horse of unknown bloodlines is a grade horse.

Genetics

Genetics is the study of genes, genetic variation, and heredity in living organisms. It is generally considered a field of biology, but it intersects frequently with many of the life sciences and is strongly linked with the study of information systems.

The father of genetics is Gregor Mendel, a late 19th-century scientist and Augustinian friar. Mendel studied "trait inheritance," patterns in the way traits are handed down from parents to offspring.

He observed that organisms (pea plants) inherit traits by way of discrete "units of inheritance." This term, still used today, is a somewhat ambiguous definition of what is referred to as a gene.

Trait inheritance and molecular inheritance mechanisms of genes are still primary principles of genetics in the 21st century, but modern genetics has expanded beyond inheritance to studying the function and behavior of genes. Gene structure and function, variation, and distribution are studied within the context of the cell, the organism (e.g. dominance), and within the context of a population. Genetics has given rise to a number of sub-fields, including epigenetics and population genetics. Organisms studied within the broad field span the domain of life, including bacteria, plants, animals, and humans.

Genetic processes work in combination with an organism's environment and experiences to influence development and behavior, often referred to as nature versus nurture. The intra- or extra-cellular environment of a cell or organism may switch gene transcription on or off. A classic example is two seeds of genetically identical corn, one placed in a temperate climate and one in an arid climate. While the average height of the two corn stalks may be genetically determined to be equal, the one in the arid climate only grows to half the height of the one in the temperate climate due to lack of water and nutrients in its environment.

History

The observation that living things inherit traits from their parents has been used since prehistoric times to improve crop plants and animals through selective breeding. The modern science of genetics, seeking to understand this process, began with the work of Gregor Mendel in the mid-19th century.

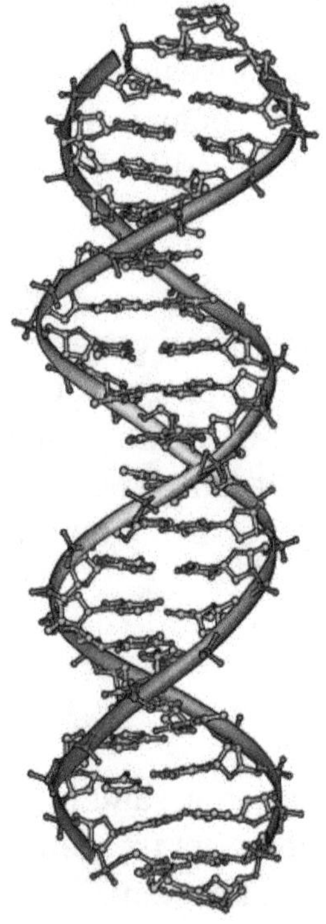

DNA, the molecular basis for biological inheritance. Each strand of DNA is a chain of nucleotides, matching each other in the center to form what look like rungs on a twisted ladder.

Prior to Mendel, Imre Festetics, a Hungarian noble, who lived in Kőszeg before Mendel, was the first who used the word "genetics." He described several rules of genetic inheritance in his work *The genetic law of the Nature* (Die genetische Gesätze der Natur, 1819). His second law is the same as what Mendel published. In his third law, he developed the basic principles of mutation (he can be considered a forerunner of Hugo de Vries.)

Other theories of inheritance preceded his work. A popular theory during Mendel's time was the concept of blending inheritance: the idea that individuals inherit a smooth blend of traits from their parents. Mendel's work provided examples where traits were definitely not blended after hybridization, showing that traits are produced by combinations of distinct genes rather than a continuous blend. Blending of traits in the progeny is now explained by the action of multiple genes with quantitative effects. Another theory that had some support at that time was the inheritance of acquired characteristics: the belief that individuals inherit traits strengthened by their parents. This theory (commonly associated with Jean-Baptiste Lamarck) is now known to be wrong—the experiences of individuals do not affect the genes they pass to their children, although evidence in the field of epigenetics has revived some aspects of Lamarck's theory. Other theories included the pangenesis of Charles Darwin (which had both acquired and inherited aspects) and Francis Galton's reformulation of pangenesis as both particulate and inherited.

Mendelian and Classical Genetics

Modern genetics started with Gregor Johann Mendel, a scientist and Augustinian friar who studied the nature of inheritance in plants. In his paper "*Versuche über Pflanzenhybriden*" ("Experiments on Plant Hybridization"), presented in 1865 to the *Naturforschender Verein* (Society for Research in Nature) in Brünn, Mendel traced the inheritance patterns of certain traits in pea plants and described them mathematically. Although this pattern of inheritance could only be observed for a few traits, Mendel's work suggested that heredity was particulate, not acquired, and that the inheritance patterns of many traits could be explained through simple rules and ratios.

The importance of Mendel's work did not gain wide understanding until the 1890s, after his death, when other scientists working on similar problems re-discovered his research. William Bateson, a proponent of Mendel's work, coined the word *genetics* in 1905. Bateson both acted as a mentor and was aided significantly by the work of female scientists from Newnham College at Cambridge, specifically the work of Becky Saunders, Nora Darwin Barlow, and Muriel Wheldale Onslow. Bateson popularized the usage of the word *genetics* to describe the study of inheritance in his inaugural address to the Third International Conference on Plant Hybridization in London, England, in 1906.

After the rediscovery of Mendel's work, scientists tried to determine which molecules in the cell were responsible for inheritance. In 1911, Thomas Hunt Morgan argued that genes are on chromosomes, based on observations of a sex-linked white eye mutation in fruit flies. In 1913, his student Alfred Sturtevant used the phenomenon of genetic linkage to show that genes are arranged linearly on the chromosome.

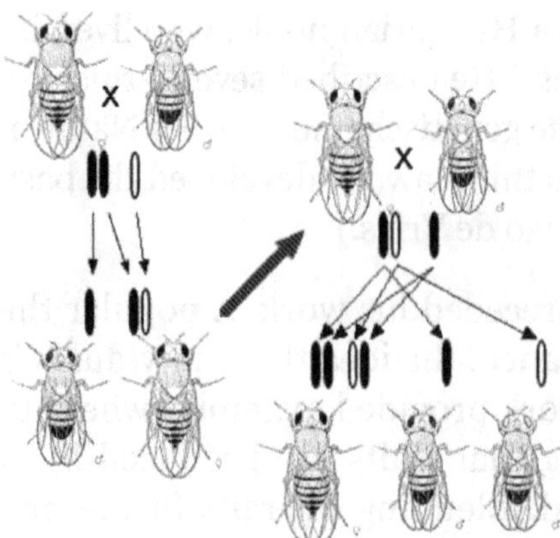

Morgan's observation of sex-linked inheritance of a mutation causing white eyes in *Drosophila* led him to the hypothesis that genes are located upon chromosomes.

Molecular Genetics

Although genes were known to exist on chromosomes, chromosomes are composed of both protein and DNA, and scientists did not know which of the two is responsible for inheritance. In 1928, Frederick Griffith discovered the phenomenon of transformation: dead bacteria could transfer genetic material to "transform" other still-living bacteria. Sixteen years later, in 1944, the Avery–MacLeod–McCarty experiment identified DNA as the molecule responsible for transformation. The role of the nucleus as the repository of genetic information in eukaryotes had been established by Hämmerling in 1943 in his work on the single celled alga *Acetabularia*. The Hershey–Chase experiment in 1952 confirmed that DNA (rather than protein) is the genetic material of the viruses that infect bacteria, providing further evidence that DNA is the molecule responsible for inheritance.

James Watson and Francis Crick determined the structure of DNA in 1953, using the X-ray crystallography work of Rosalind Franklin and Maurice Wilkins that indicated DNA has a helical structure (i.e., shaped like a corkscrew). Their double-helix model had two strands of DNA with the nucleotides pointing inward, each matching a complementary nucleotide on the other strand to form what look like rungs on a twisted ladder. This structure showed that genetic information exists in the sequence of nucleotides on each strand of DNA. The structure also suggested a simple method for replication: if the strands are separated, new partner strands can be reconstructed for each based on the sequence of the old strand. This property is what gives DNA its semi-conservative nature where one strand of new DNA is from an original parent strand.

Although the structure of DNA showed how inheritance works, it was still not known how DNA influences the behavior of cells. In the following years, scientists tried to understand how DNA controls the process of protein production. It was discovered that the cell uses DNA as a template to create matching messenger RNA, molecules with nucleotides very similar to DNA. The nucleotide sequence of a messenger RNA is used to create an amino acid sequence in protein; this translation between nucleotide sequences and amino acid sequences is known as the genetic code.

With the newfound molecular understanding of inheritance came an explosion of research. A notable theory arose from Tomoko Ohta in 1973 with her amendment to the neutral theory of molecular evolution through publishing the nearly neutral theory of molecular evolution. In this theory, Ohta stressed the importance of natural selection and the environment to the rate at which genetic evolution occurs. One important development was chain-termination DNA sequencing in 1977 by Frederick Sanger. This technology allows scientists to read the nucleotide sequence of a DNA molecule. In 1983, Kary Banks Mullis developed the polymerase chain reaction, providing a quick way to isolate and amplify a specific section of DNA from a mixture. The efforts of the Human Genome Project, Department of Energy, NIH, and parallel private efforts by Celera Genomics led to the sequencing of the human genome in 2003.

Features of Inheritance

Discrete Inheritance and Mendel's Laws

At its most fundamental level, inheritance in organisms occurs by passing discrete heritable units, called genes, from parents to offspring. This property was first observed by Gregor Mendel, who studied the segregation of heritable traits in pea plants. In his experiments studying the trait for flower color, Mendel observed that the flowers of each pea plant were either purple or white—but never an intermediate between the two colors. These different, discrete versions of the same gene are called alleles.

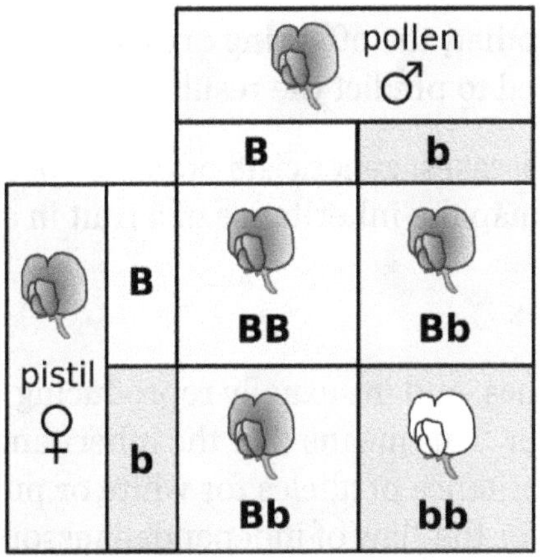

A Punnett square depicting a cross between two pea plants heterozygous for purple (B) and white (b) blossoms.

In the case of the pea, which is a diploid species, each individual plant has two copies of each gene, one copy inherited from each parent. Many species, including humans, have this pattern of inheritance. Diploid organisms with two copies of the same allele of a given gene are called homozygous at that gene locus, while organisms with two different alleles of a given gene are called heterozygous.

The set of alleles for a given organism is called its genotype, while the observable traits of the organism are called its phenotype. When organisms are heterozygous at a gene, often one allele is called dominant as its qualities dominate the phenotype of the organism, while the other allele is called recessive as its qualities recede and are not observed. Some alleles do not have complete dominance and instead have incomplete dominance by expressing an intermediate phenotype, or codominance by expressing both alleles at once.

When a pair of organisms reproduce sexually, their offspring randomly inherit one of the two alleles from each parent. These observations of discrete inheritance and the segregation of alleles are collectively known as Mendel's first law or the Law of Segregation.

Notation and Diagrams

Geneticists use diagrams and symbols to describe inheritance. A gene is represented by one or a few letters. Often a "+" symbol is used to mark the usual, non-mutant allele for a gene.

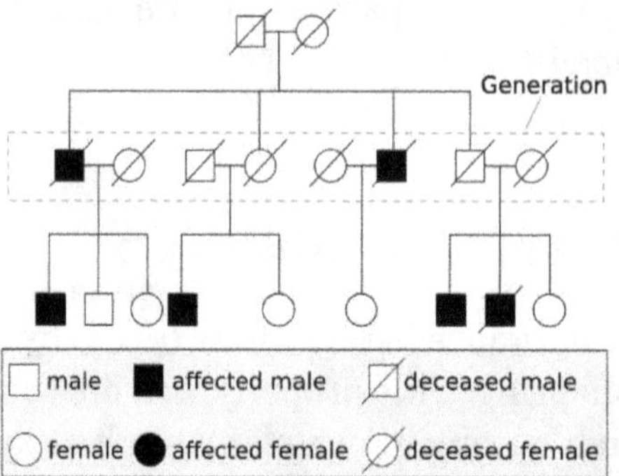

Genetic pedigree charts help track the inheritance patterns of traits.

In fertilization and breeding experiments (and especially when discussing Mendel's laws) the parents are referred to as the "P" generation and the offspring as the "F1" (first filial) generation. When the F1 offspring mate with each other, the offspring are called the "F2" (second filial) generation. One of the common diagrams used to predict the result of cross-breeding is the Punnett square.

When studying human genetic diseases, geneticists often use pedigree charts to represent the inheritance of traits. These charts map the inheritance of a trait in a family tree.

Multiple Gene Interactions

Organisms have thousands of genes, and in sexually reproducing organisms these genes generally assort independently of each other. This means that the inheritance of an allele for yellow or green pea color is unrelated to the inheritance of alleles for white or purple flowers. This phenomenon, known as "Mendel's second law" or the "law of independent assortment," means that the alleles of different genes get shuffled between parents to form offspring with many different combinations.

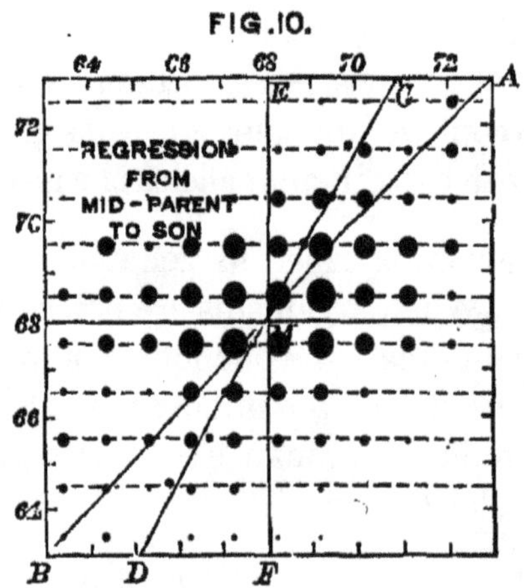

Human height is a trait with complex genetic causes. Francis Galton's data from 1889 shows the relationship between offspring height as a function of mean parent height.

Often different genes can interact in a way that influences the same trait. In the Blue-eyed Mary (*Omphalodes verna*), for example, there exists a gene with alleles that determine the color of flowers: blue or magenta. Another gene, however, controls whether the flowers have color at all or are white. When a plant has two copies of this white allele, its flowers are white—regardless of whether the first gene has blue or magenta alleles. This interaction between genes is called epistasis, with the second gene epistatic to the first.

Many traits are not discrete features (e.g. purple or white flowers) but are instead continuous features (e.g. human height and skin color). These complex traits are products of many genes. The influence of these genes is mediated, to varying degrees, by the environment an organism has experienced. The degree to which an organism's genes contribute to a complex trait is called heritability. Measurement of the heritability of a trait is relative—in a more variable environment, the environment has a bigger influence on the total variation of the trait. For example, human height is a trait with complex causes. It has a heritability of 89% in the United States. In Nigeria, however, where people experience a more variable access to good nutrition and health care, height has a heritability of only 62%.

Molecular Basis for Inheritance

DNA and Chromosomes

The molecular basis for genes is deoxyribonucleic acid (DNA). DNA is composed of a chain of nucleotides, of which there are four types: adenine (A), cytosine (C), guanine (G), and thymine (T). Genetic information exists in the sequence of these nucleotides, and genes exist as stretches of sequence along the DNA chain. Viruses are the only exception to this rule—sometimes viruses use the very similar molecule RNA instead of DNA as their genetic material. Viruses cannot reproduce without a host and are unaffected by many genetic processes, so tend not to be considered living organisms.

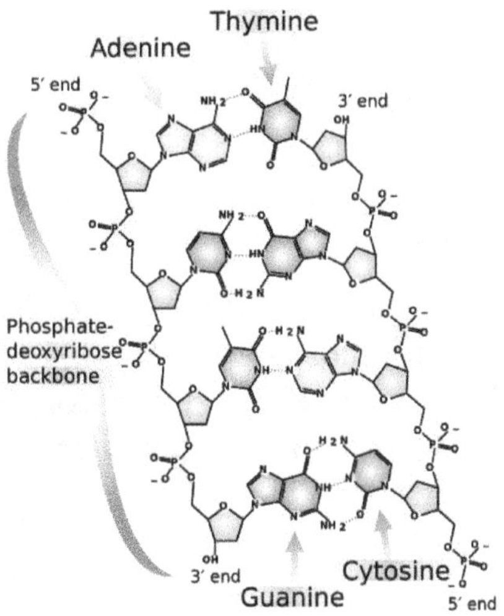

The molecular structure of DNA. Bases pair through the arrangement of hydrogen bonding between the strands.

DNA normally exists as a double-stranded molecule, coiled into the shape of a double helix. Each nucleotide in DNA preferentially pairs with its partner nucleotide on the opposite strand: A pairs with T, and C pairs with G. Thus, in its two-stranded form, each strand effectively contains all necessary information, redundant with its partner strand. This structure of DNA is the physical basis for inheritance: DNA replication duplicates the genetic information by splitting the strands and using each strand as a template for synthesis of a new partner strand.

Genes are arranged linearly along long chains of DNA base-pair sequences. In bacteria, each cell usually contains a single circular genophore, while eukaryotic organisms (such as plants and animals) have their DNA arranged in multiple linear chromosomes. These DNA strands are often extremely long; the largest human chromosome, for example, is about 247 million base pairs in length. The DNA of a chromosome is associated with structural proteins that organize, compact, and control access to the DNA, forming a material called chromatin; in eukaryotes, chromatin is usually composed of nucleosomes, segments of DNA wound around cores of histone proteins. The full set of hereditary material in an organism (usually the combined DNA sequences of all chromosomes) is called the genome.

While haploid organisms have only one copy of each chromosome, most animals and many plants are diploid, containing two of each chromosome and thus two copies of every gene. The two alleles for a gene are located on identical loci of the two homologous chromosomes, each allele inherited from a different parent.

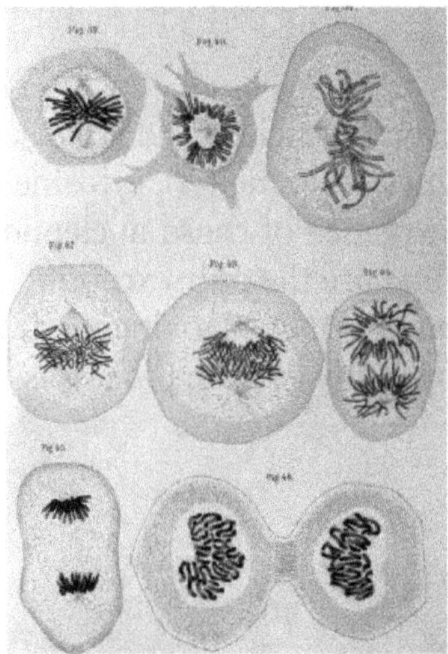

Walther Flemming's 1882 diagram of eukaryotic cell division. Chromosomes are copied, condensed, and organized. Then, as the cell divides, chromosome copies separate into the daughter cells.

Many species have so-called sex chromosomes that determine the gender of each organism. In humans and many other animals, the Y chromosome contains the gene that triggers the development of the specifically male characteristics. In evolution, this chromosome has lost most of its content and also most of its genes, while the X chromosome is similar to the other chromosomes and contains many genes. The X and Y chromosomes form a strongly heterogeneous pair.

Reproduction

When cells divide, their full genome is copied and each daughter cell inherits one copy. This process, called mitosis, is the simplest form of reproduction and is the basis for asexual reproduction. Asexual reproduction can also occur in multicellular organisms, producing offspring that inherit their genome from a single parent. Offspring that are genetically identical to their parents are called clones.

Eukaryotic organisms often use sexual reproduction to generate offspring that contain a mixture of genetic material inherited from two different parents. The process of sexual reproduction alternates between forms that contain single copies of the genome (haploid) and double copies (diploid). Haploid cells fuse and combine genetic material to create a diploid cell with paired chromosomes. Diploid organisms form haploids by dividing, without replicating their DNA, to create daughter cells that randomly inherit one of each pair of chromosomes. Most animals and many plants are diploid for most of their lifespan, with the haploid form reduced to single cell gametes such as sperm or eggs.

Although they do not use the haploid/diploid method of sexual reproduction, bacteria have many methods of acquiring new genetic information. Some bacteria can undergo conjugation, transferring a small circular piece of DNA to another bacterium. Bacteria can also take up raw DNA fragments found in the environment and integrate them into their genomes, a phenomenon known as transformation. These processes result in horizontal gene transfer, transmitting fragments of genetic information between organisms that would be otherwise unrelated.

Recombination and Genetic Linkage

The diploid nature of chromosomes allows for genes on different chromosomes to assort independently or be separated from their homologous pair during sexual reproduction wherein haploid gametes are formed. In this way new combinations of genes can occur in the offspring of a mating pair. Genes on the same chromosome would theoretically never recombine. However, they do, via the cellular process of chromosomal crossover. During crossover, chromosomes exchange stretches of DNA, effectively shuffling the gene alleles between the chromosomes. This process of chromosomal crossover generally occurs during meiosis, a series of cell divisions that creates haploid cells.

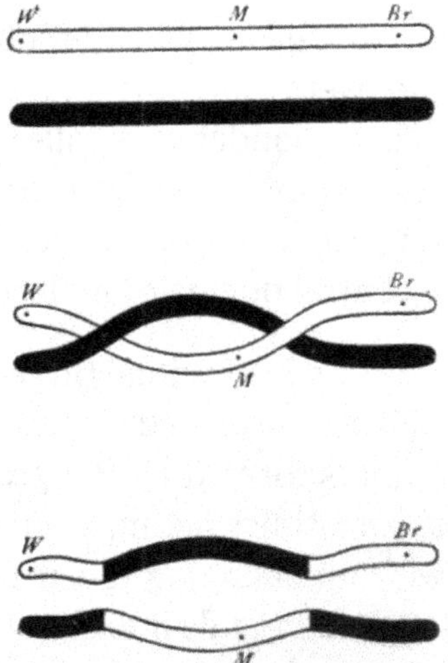

Thomas Hunt Morgan's 1916 illustration of a double crossover between chromosomes.

The first cytological demonstration of crossing over was performed by Harriet Creighton and Barbara McClintock in 1931. Their research and experiments on corn provided cytological evidence for the genetic theory that linked genes on paired chromosomes do in fact exchange places from one homolog to the other.

The probability of chromosomal crossover occurring between two given points on the chromosome is related to the distance between the points. For an arbitrarily long distance, the probability of crossover is high enough that the inheritance of the genes is effectively uncorrelated. For genes that are closer together, however, the lower probability of crossover means that the genes demonstrate genetic linkage; alleles for the two genes tend to be inherited together. The amounts of linkage between a series of genes can be combined to form a linear linkage map that roughly describes the arrangement of the genes along the chromosome.

Gene Expression

Genetic Code

Genes generally express their functional effect through the production of proteins, which are complex molecules responsible for most functions in the cell. Proteins are made up of one or more polypeptide chains, each of which is composed of a sequence of amino acids, and the DNA sequence of a gene (through an RNA intermediate) is used to produce a specific amino acid sequence. This process begins with the production of an RNA molecule with a sequence matching the gene's DNA sequence, a process called transcription.

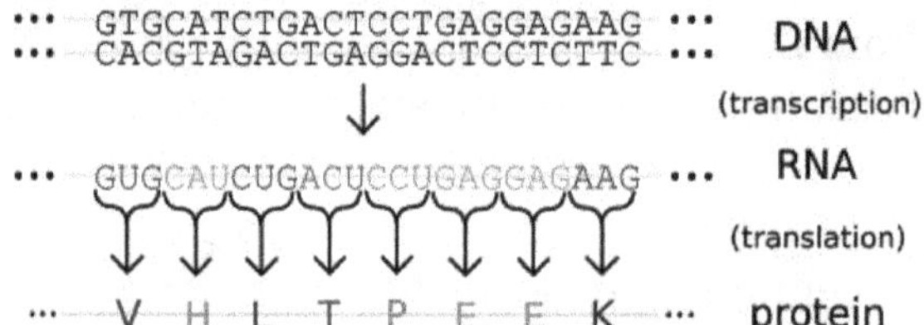

The genetic code: Using a triplet code, DNA, through a messenger RNA intermediary, specifies a protein.

This messenger RNA molecule is then used to produce a corresponding amino acid sequence through a process called translation. Each group of three nucleotides in the sequence, called a codon, corresponds either to one of the twenty possible amino acids in a protein or an instruction to end the amino acid sequence; this correspondence is called the genetic code. The flow of information is unidirectional: information is transferred from nucleotide sequences into the amino acid sequence of proteins, but it never transfers from protein back into the sequence of DNA—a phenomenon Francis Crick called the central dogma of molecular biology.

The specific sequence of amino acids results in a unique three-dimensional structure for that protein, and the three-dimensional structures of proteins are related to their functions. Some are simple structural molecules, like the fibers formed by the protein collagen. Proteins can bind to other proteins and simple molecules, sometimes acting as enzymes by facilitating chemical reactions within the bound molecules (without changing the structure of the protein itself). Protein structure is dynamic; the protein hemoglobin bends into slightly different forms as it facilitates the capture, transport, and release of oxygen molecules within mammalian blood.

A single nucleotide difference within DNA can cause a change in the amino acid sequence of a protein. Because protein structures are the result of their amino acid sequences, some changes can dramatically change the properties of a protein by destabilizing the structure or changing the surface of the protein in a way that changes its interaction with other proteins and molecules. For example, sickle-cell anemia is a human genetic disease that results from a single base difference within the coding region for the β-globin section of hemoglobin, causing a single amino acid change that changes hemoglobin's physical properties. Sickle-cell versions of hemoglobin stick to themselves, stacking to form fibers that distort the shape of red blood cells carrying the protein. These sickle-shaped cells no longer flow smoothly through blood vessels, having a tendency to clog or degrade, causing the medical problems associated with this disease.

Some DNA sequences are transcribed into RNA but are not translated into protein products—such RNA molecules are called non-coding RNA. In some cases, these products fold into structures which are involved in critical cell functions (e.g. ribosomal RNA and transfer RNA). RNA can also have regulatory effects through hybridization interactions with other RNA molecules (e.g. microRNA).

Nature and Nurture

Although genes contain all the information an organism uses to function, the environment plays an important role in determining the ultimate phenotypes an organism displays. This is the complementary relationship often referred to as "nature and nurture." The phenotype of an organism depends on the interaction of genes and the environment. An interesting example is the coat coloration of the Siamese cat. In this case, the body temperature of the cat plays the role of the environment. The cat's genes code for dark hair, thus the hair-producing cells in the cat make cellular proteins resulting in dark hair. But these dark hair-producing proteins are sensitive to temperature (i.e. have a mutation causing temperature-sensitivity) and denature in higher-temperature environments, failing to produce dark-hair pigment in areas where the cat has a higher body temperature. In a low-temperature environment, however, the protein's structure is stable and produces dark-hair pigment normally. The protein remains functional in areas of skin that are colder – such as its legs, ears, tail and face – so the cat has dark-hair at its extremities.

Siamese cats have a temperature-sensitive pigment-production mutation.

Environment plays a major role in effects of the human genetic disease phenylketonuria. The mutation that causes phenylketonuria disrupts the ability of the body to break down the amino acid phenylalanine, causing a toxic build-up of an intermediate molecule that, in turn, causes severe symptoms of progressive intellectual disability and seizures. However, if someone with the phenylketonuria mutation follows a strict diet that avoids this amino acid, they remain normal and healthy.

A popular method for determining how genes and environment ("nature and nurture") contribute to a phenotype involves studying identical and fraternal twins, or other siblings of multiple births. Because identical siblings come from the same zygote, they are genetically the same. Fraternal twins are as genetically different from one another as normal siblings. By comparing how often a certain disorder occurs in a pair of identical twins to how often it occurs in a pair of fraternal twins, scientists can determine whether that disorder is caused by genetic or postnatal environmental factors – whether it has "nature" or "nurture" causes. One famous example is the multiple birth study of the Genain quadruplets, who were identical quadruplets all diagnosed with schizophrenia. However such tests cannot separate genetic factors from environmental factors affecting fetal development.

Gene Regulation

The genome of a given organism contains thousands of genes, but not all these genes need to be active at any given moment. A gene is expressed when it is being transcribed into mRNA and there exist many cellular methods of controlling the expression of genes such that proteins are produced only when needed by the cell. Transcription factors are regulatory proteins that bind to DNA, either promoting or inhibiting the transcription of a gene. Within the genome of *Escherichia coli* bacteria, for example, there exists a series of genes necessary for the synthesis of the amino acid tryptophan. However, when tryptophan is already available to the cell, these genes for tryptophan synthesis are no longer needed. The presence of tryptophan directly affects the activity of the genes—tryptophan molecules bind to the tryptophan repressor (a transcription factor), changing the repressor's structure such that the repressor binds to the genes. The tryptophan repressor blocks the transcription and expression of the genes, thereby creating negative feedback regulation of the tryptophan synthesis process.

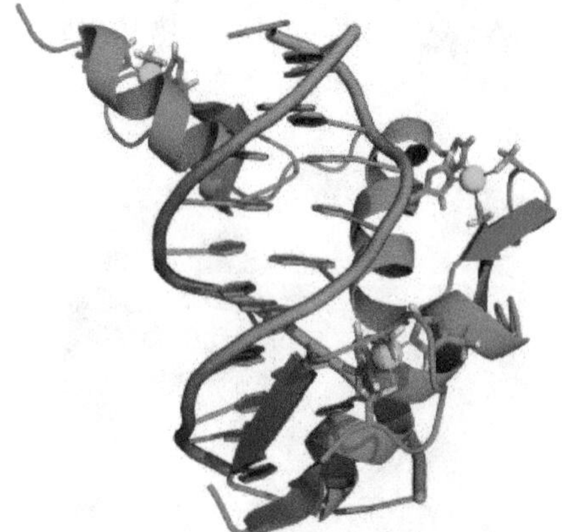

Transcription factors bind to DNA, influencing the transcription of associated genes.

Differences in gene expression are especially clear within multicellular organisms, where cells all contain the same genome but have very different structures and behaviors due to the expression of different sets of genes. All the cells in a multicellular organism derive from a single cell, differentiating into variant cell types in response to external and intercellular signals and gradually establishing different patterns of gene expression to create different behaviors. As no single gene is responsible for the development of structures within multicellular organisms, these patterns arise from the complex interactions between many cells.

Within eukaryotes, there exist structural features of chromatin that influence the transcription of genes, often in the form of modifications to DNA and chromatin that are stably inherited by daughter cells. These features are called "epigenetic" because they exist "on top" of the DNA sequence and retain inheritance from one cell generation to the next. Because of epigenetic features, different cell types grown within the same medium can retain very different properties. Although epigenetic features are generally dynamic over the course of development, some, like the phenomenon of paramutation, have multigenerational inheritance and exist as rare exceptions to the general rule of DNA as the basis for inheritance.

Genetic Change

Mutations

During the process of DNA replication, errors occasionally occur in the polymerization of the second strand. These errors, called mutations, can affect the phenotype of an organism, especially if they occur within the protein coding sequence of a gene. Error rates are usually very low—1 error in every 10–100 million bases—due to the "proofreading" ability of DNA polymerases. Processes that increase the rate of changes in DNA are called mutagenic: mutagenic chemicals promote errors in DNA replication, often by interfering with the structure of base-pairing, while UV radiation induces mutations by causing damage to the DNA structure. Chemical damage to DNA occurs naturally as well and cells use DNA repair mechanisms to repair mismatches and breaks. The repair does not, however, always restore the original sequence.

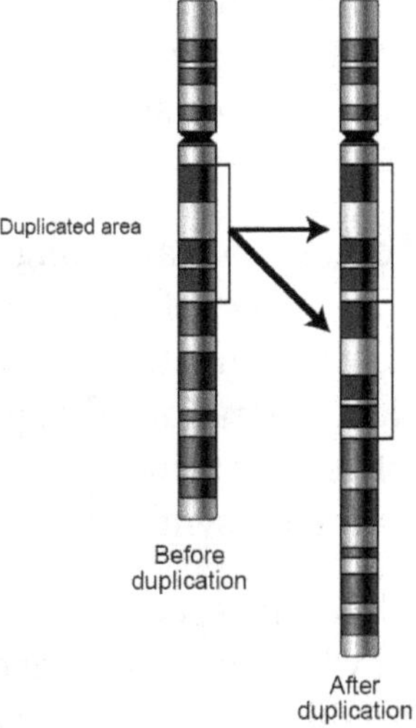

Gene duplication allows diversification by providing redundancy: one gene can mutate and lose its original function without harming the organism.

In organisms that use chromosomal crossover to exchange DNA and recombine genes, errors in alignment during meiosis can also cause mutations. Errors in crossover are especially likely when similar sequences cause partner chromosomes to adopt a mistaken alignment; this makes some regions in genomes more prone to mutating in this way. These errors create large structural changes in DNA sequence – duplications, inversions, deletions of entire regions – or the accidental exchange of whole parts of sequences between different chromosomes (chromosomal translocation).

Natural Selection and Evolution

Mutations alter an organism's genotype and occasionally this causes different phenotypes to appear. Most mutations have little effect on an organism's phenotype, health, or reproductive fitness. Mutations that do have an effect are usually detrimental, but occasionally some can be beneficial. Studies in the fly *Drosophila melanogaster* suggest that if a mutation changes a protein produced by a gene, about 70 percent of these mutations will be harmful with the remainder being either neutral or weakly beneficial.

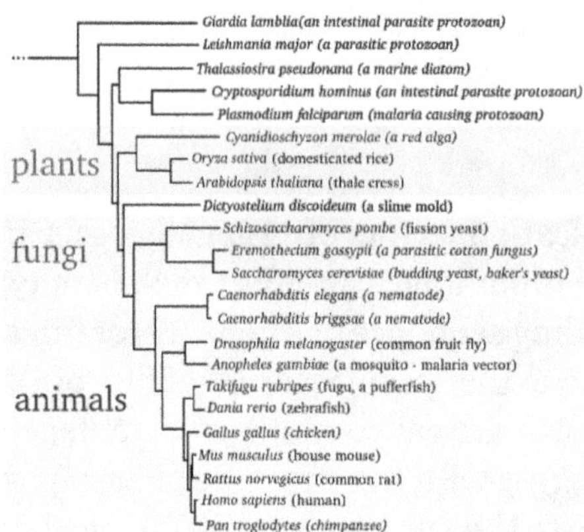

An evolutionary tree of eukaryotic organisms, constructed by the comparison of several orthologous gene sequences.

Population genetics studies the distribution of genetic differences within populations and how these distributions change over time. Changes in the frequency of an allele in a population are mainly influenced by natural selection, where a given allele provides a selective or reproductive advantage to the organism, as well as other factors such as mutation, genetic drift, genetic draft, artificial selection and migration.

Over many generations, the genomes of organisms can change significantly, resulting in evolution. In the process called adaptation, selection for beneficial mutations can cause a species to evolve into forms better able to survive in their environment. New species are formed through the process of speciation, often caused by geographical separations that prevent populations from exchanging genes with each other. The application of genetic principles to the study of population biology and evolution is known as the "modern evolutionary synthesis."

By comparing the homology between different species' genomes, it is possible to calculate the evolutionary distance between them and when they may have diverged. Genetic comparisons are generally considered a more accurate method of characterizing the relatedness between species than the comparison of phenotypic characteristics. The evolutionary distances between species can be used to form evolutionary trees; these trees represent the common descent and divergence of species over time, although they do not show the transfer of genetic material between unrelated species (known as horizontal gene transfer and most common in bacteria).

Model Organisms

Although geneticists originally studied inheritance in a wide range of organisms, researchers began to specialize in studying the genetics of a particular subset of organisms. The fact that significant research already existed for a given organism would encourage new researchers to choose it for further study, and so eventually a few model organisms became the basis for most genetics research. Common research topics in model organism genetics include the study of gene regulation and the involvement of genes in development and cancer.

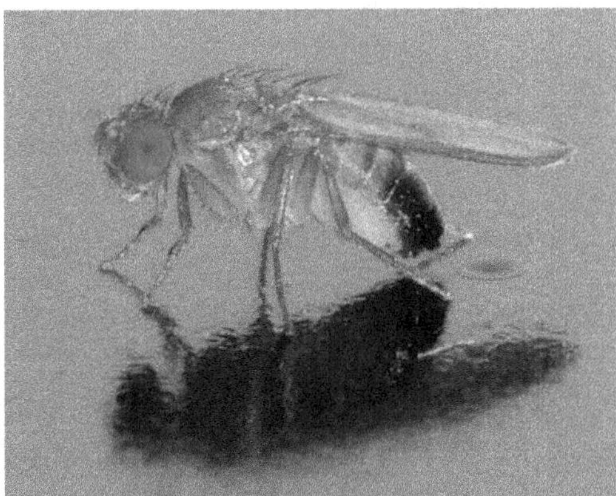

The common fruit fly (*Drosophila melanogaster*) is a popular model organism in genetics research.

Organisms were chosen, in part, for convenience—short generation times and easy genetic manipulation made some organisms popular genetics research tools. Widely used model organisms include the gut bacterium *Escherichia coli*, the plant *Arabidopsis thaliana*, baker's yeast (*Saccharomyces cerevisiae*), the nematode *Caenorhabditis elegans*, the common fruit fly (*Drosophila melanogaster*), and the common house mouse (*Mus musculus*).

Medicine

Medical genetics seeks to understand how genetic variation relates to human health and disease. When searching for an unknown gene that may be involved in a disease, researchers commonly use genetic linkage and genetic pedigree charts to find the location on the genome associated with the disease. At the population level, researchers take advantage of Mendelian randomization to look for locations in the genome that are associated with diseases, a method especially useful for multigenic traits not clearly defined by a single gene. Once a candidate gene is found, further research is often done on the corresponding (or homologous) genes of model organisms. In addition to studying genetic diseases, the increased availability of genotyping methods has led to the field of pharmacogenetics: the study of how genotype can affect drug responses.

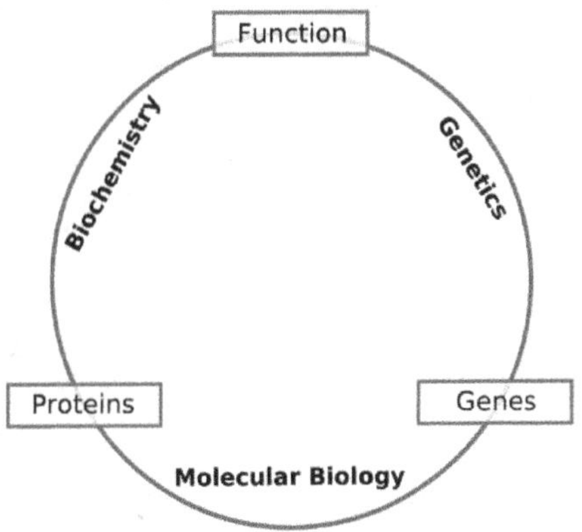

Schematic relationship between biochemistry, genetics and molecular biology.

Individuals differ in their inherited tendency to develop cancer, and cancer is a genetic disease. The process of cancer development in the body is a combination of events. Mutations occasionally occur within cells in the body as they divide. Although these mutations will not be inherited by any offspring, they can affect the behavior of cells, sometimes causing them to grow and divide more frequently. There are biological mechanisms that attempt to stop this process; signals are given to inappropriately dividing cells that should trigger cell death, but sometimes additional mutations occur that cause cells to ignore these messages. An internal process of natural selection occurs within the body and eventually mutations accumulate within cells to promote their own growth, creating a cancerous tumor that grows and invades various tissues of the body.

Normally, a cell divides only in response to signals called growth factors and stops growing once in contact with surrounding cells and in response to growth-inhibitory signals. It usually then divides a limited number of times and dies, staying within the epithelium where it is unable to migrate to other organs. To become a cancer cell, a cell has to accumulate mutations in a number of genes (three to seven) that allow it to bypass this regulation: it no longer needs growth factors to divide, continues growing when making contact to neighbor cells, ignores inhibitory signals, keeps growing indefinitely and is immortal, escapes from the epithelium and ultimately may be able to escape from the primary tumor, cross the endothelium of a blood vessel, be transported by the bloodstream and colonize a new organ, forming deadly metastasis. Although there are some genetic predispositions in a small fraction of cancers, the major fraction is due to a set of new genetic mutations that originally appear and accumulate in one or a small number of cells that will divide to form the tumor and are not transmitted to the progeny (somatic mutations). The most frequent mutations are a loss of function of p53 protein, a tumor suppressor, or in the p53 pathway, and gain of function mutations in the Ras proteins, or in other oncogenes.

Research Methods

DNA can be manipulated in the laboratory. Restriction enzymes are commonly used enzymes that cut DNA at specific sequences, producing predictable fragments of DNA. DNA fragments can be visualized through use of gel electrophoresis, which separates fragments according to their length.

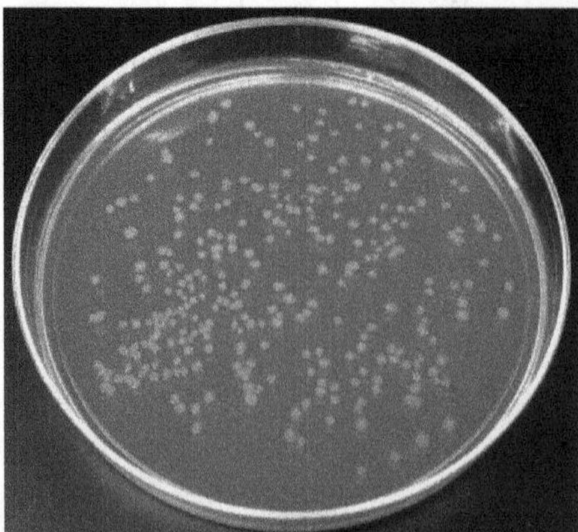

Colonies of *E. coli* produced by cellular cloning. A similar methodology is often used in molecular cloning.

The use of ligation enzymes allows DNA fragments to be connected. By binding ("ligating") fragments of DNA together from different sources, researchers can create recombinant DNA, the DNA often associated with genetically modified organisms. Recombinant DNA is commonly used in the context of plasmids: short circular DNA molecules with a few genes on them. In the process known as molecular cloning, researchers can amplify the DNA fragments by inserting plasmids into bacteria and then culturing them on plates of agar (to isolate clones of bacteria cells). ("Cloning" can also refer to the various means of creating cloned ("clonal") organisms.)

DNA can also be amplified using a procedure called the polymerase chain reaction (PCR). By using specific short sequences of DNA, PCR can isolate and exponentially amplify a targeted region of DNA. Because it can amplify from extremely small amounts of DNA, PCR is also often used to detect the presence of specific DNA sequences.

DNA Sequencing and Genomics

DNA sequencing, one of the most fundamental technologies developed to study genetics, allows researchers to determine the sequence of nucleotides in DNA fragments. The technique of chain-termination sequencing, developed in 1977 by a team led by Frederick Sanger, is still routinely used to sequence DNA fragments. Using this technology, researchers have been able to study the molecular sequences associated with many human diseases.

As sequencing has become less expensive, researchers have sequenced the genomes of many organisms using a process called genome assembly, which utilizes computational tools to stitch together sequences from many different fragments. These technologies were used to sequence the human genome in the Human Genome Project completed in 2003. New high-throughput sequencing technologies are dramatically lowering the cost of DNA sequencing, with many researchers hoping to bring the cost of resequencing a human genome down to a thousand dollars.

Next generation sequencing (or high-throughput sequencing) came about due to the ever-increasing demand for low-cost sequencing. These sequencing technologies allow the production of potentially millions of sequences concurrently. The large amount of sequence data available has created the field of genomics, research that uses computational tools to search for and analyze patterns in the full genomes of organisms. Genomics can also be considered a subfield of bioinformatics, which uses computational approaches to analyze large sets of biological data. A common problem to these fields of research is how to manage and share data that deals with human subject and personally identifiable information.

Society and Culture

On 19 March 2015, a leading group of biologists urged a worldwide ban on clinical use of methods, particularly the use of CRISPR and zinc finger, to edit the human genome in a way that can be inherited. In April 2015, Chinese researchers reported results of basic research to edit the DNA of non-viable human embryos using CRISPR.

References

- Judson, Horace (1979). The Eighth Day of Creation: Makers of the Revolution in Biology. Cold Spring Harbor Laboratory Press. pp. 51–169. ISBN 0-87969-477-7.
- Frederick Betz (2010). Managing Science: Methodology and Organization of Research. Springer. p. 76. ISBN 978-1-4419-7488-4.
- Griffiths, Anthony J. F.; Miller, Jeffrey H.; Suzuki, David T.; Lewontin, Richard C.; Gelbart, eds. (2000). "Patterns of Inheritance: Introduction". An Introduction to Genetic Analysis (7th ed.). New York: W. H. Freeman. ISBN 0-7167-3520-2.
- Griffiths, Anthony J. F.; Miller, Jeffrey H.; Suzuki, David T.; Lewontin, Richard C.; Gelbart, eds. (2000). "Mendel's experiments". An Introduction to Genetic Analysis (7th ed.). New York: W. H. Freeman. ISBN 0-7167-3520-2.
- Griffiths, Anthony J. F.; Miller, Jeffrey H.; Suzuki, David T.; Lewontin, Richard C.; Gelbart, eds. (2000). "Mendelian genetics in eukaryotic life cycles". An Introduction to Genetic Analysis (7th ed.). New York: W. H. Freeman. ISBN 0-7167-3520-2.
- Griffiths, Anthony J. F.; Miller, Jeffrey H.; Suzuki, David T.; Lewontin, Richard C.; Gelbart, eds. (2000). "Interactions between the alleles of one gene". An Introduction to Genetic Analysis (7th ed.). New York: W. H. Freeman. ISBN 0-7167-3520-2.
- Griffiths, Anthony J. F.; Miller, Jeffrey H.; Suzuki, David T.; Lewontin, Richard C.; Gelbart, eds. (2000). "Human Genetics". An Introduction to Genetic Analysis (7th ed.). New York: W. H. Freeman. ISBN 0-7167-3520-2.
- Griffiths, Anthony J. F.; Miller, Jeffrey H.; Suzuki, David T.; Lewontin, Richard C.; Gelbart, eds. (2000). "Gene interaction and modified dihybrid ratios". An Introduction to Genetic Analysis (7th ed.). New York: W. H. Freeman. ISBN 0-7167-3520-2
- Griffiths, Anthony J. F.; Miller, Jeffrey H.; Suzuki, David T.; Lewontin, Richard C.; Gelbart, eds. (2000). "Quantifying heritability". An Introduction to Genetic Analysis (7th ed.). New York: W. H. Freeman. ISBN 0-7167-3520-2.
- Griffiths, Anthony J. F.; Miller, Jeffrey H.; Suzuki, David T.; Lewontin, Richard C.; Gelbart, eds. (2000). "Mechanism of DNA Replication". An Introduction to Genetic Analysis (7th ed.). New York: W. H. Freeman. ISBN 0-7167-3520-2.
- Griffiths, Anthony J. F.; Miller, Jeffrey H.; Suzuki, David T.; Lewontin, Richard C.; Gelbart, eds. (2000). "Sex chromosomes and sex-linked inheritance". An Introduction to Genetic Analysis (7th ed.). New York: W. H. Freeman. ISBN 0-7167-3520-2.
- Griffiths, Anthony J. F.; Miller, Jeffrey H.; Suzuki, David T.; Lewontin, Richard C.; Gelbart, eds. (2000). "Bacterial conjugation". An Introduction to Genetic Analysis (7th ed.). New York: W. H. Freeman. ISBN 0-7167-3520-2.
- Griffiths, Anthony J. F.; Miller, Jeffrey H.; Suzuki, David T.; Lewontin, Richard C.; Gelbart, eds. (2000). "Bacterial transformation". An Introduction to Genetic Analysis (7th ed.). New York: W. H. Freeman. ISBN 0-7167-3520-2.
- Griffiths, Anthony J. F.; Miller, Jeffrey H.; Suzuki, David T.; Lewontin, Richard C.; Gelbart, eds. (2000). "Nature of crossing-over". An Introduction to Genetic Analysis (7th ed.). New York: W. H. Freeman. ISBN 0-7167-3520-2.
- Jack E. Staub (1994). Crossover: Concepts and Applications in Genetics, Evolution, and Breeding. University of Wisconsin Press. p. 55. ISBN 978-0-299-13564-5.
- Griffiths, Anthony J. F.; Miller, Jeffrey H.; Suzuki, David T.; Lewontin, Richard C.; Gelbart, eds. (2000). "Linkage maps". An Introduction to Genetic Analysis (7th ed.). New York: W. H. Freeman. ISBN 0-7167-3520-2.
- For example, Ridley, M. (2003). Nature via Nurture: Genes, Experience and What Makes Us Human. Fourth

Estate. p. 73. ISBN 978-1-84115-745-0.

- Griffiths, Anthony J. F.; Miller, Jeffrey H.; Suzuki, David T.; Lewontin, Richard C.; Gelbart, eds. (2000). "Spontaneous mutations". An Introduction to Genetic Analysis (7th ed.). New York: W. H. Freeman. ISBN 0-7167-3520-2.
- Griffiths, Anthony J. F.; Miller, Jeffrey H.; Suzuki, David T.; Lewontin, Richard C.; Gelbart, eds. (2000). "Induced mutations". An Introduction to Genetic Analysis (7th ed.). New York: W. H. Freeman. ISBN 0-7167-3520-2.
- Griffiths, Anthony J. F.; Miller, Jeffrey H.; Suzuki, David T.; Lewontin, Richard C.; Gelbart, eds. (2000). "Chromosome Mutation I: Changes in Chromosome Structure: Introduction". An Introduction to Genetic Analysis (7th ed.). New York: W. H. Freeman. ISBN 0-7167-3520-2.
- Mike Calver; Alan Lymbery; Jennifer McComb; Mike Bamford (2009). Environmental Biology. Cambridge University Press. p. 118. ISBN 978-0-521-67982-4.
- Griffiths, Anthony J. F.; Miller, Jeffrey H.; Suzuki, David T.; Lewontin, Richard C.; Gelbart, eds. (2000). "Variation and its modulation". An Introduction to Genetic Analysis (7th ed.). New York: W. H. Freeman. ISBN 0-7167-3520-2.
- Griffiths, Anthony J. F.; Miller, Jeffrey H.; Suzuki, David T.; Lewontin, Richard C.; Gelbart, eds. (2000). "Selection". An Introduction to Genetic Analysis (7th ed.). New York: W. H. Freeman. ISBN 0-7167-3520-2.
- Griffiths, Anthony J. F.; Miller, Jeffrey H.; Suzuki, David T.; Lewontin, Richard C.; Gelbart, eds. (2000). "Random events". An Introduction to Genetic Analysis (7th ed.). New York: W. H. Freeman. ISBN 0-7167-3520-2.
- Brown TA (2002). "Section 2, Chapter 6: 6.1. The Methodology for DNA Sequencing". Genomes 2 (2nd ed.). Oxford: Bios. ISBN 1-85996-228-9.

Key Concepts of Genetics

The chapter closely examines the key concepts of genetics to provide an extensive understanding of the subject. Some of the key concepts discussed in this chapter are chromosomes, genetic diversity, mutation, DNA and RNA. A chromosome is a structure that contains the DNA of a living organism whereas genetic diversity is the sum of all the genetic characteristics in the genetic nature of an organism. The text strategically encompasses and incorporates the major key concepts of genetics, providing a complete understanding.

Chromosome

A chromosome is a packaged and organized structure containing most of the DNA of a living organism. DNA is not usually found on its own, but rather is structured in long strands that are wrapped around protein complexes called nucleosomes that consist of proteins called histones. The DNA in chromosomes serves as the source for transcription. Most eukaryotic cells have a set of chromosomes (46 in humans) with the genetic material spread among them.

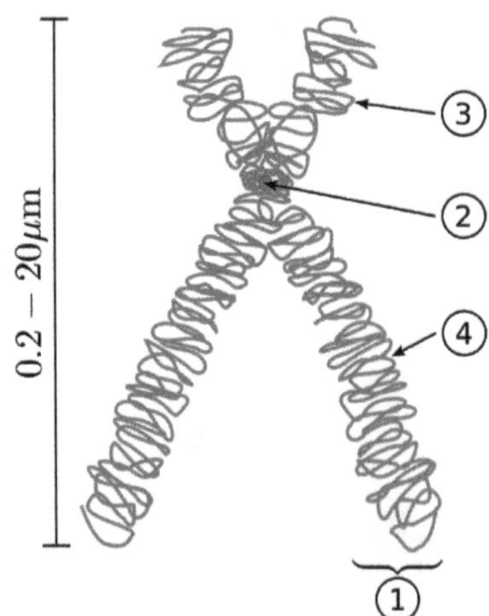

Diagram of a replicated and condensed metaphase eukaryotic chromosome. (1) Chromatid – one of the two identical parts of the chromosome after S phase. (2) Centromere – the point where the two chromatids touch. (3) Short arm. (4) Long arm.

During most of the duration of the cell cycle, a chromosome consists of one long (its width to length ratio is about 1:65,000,000) double-helix DNA molecule (with associated proteins). During S phase, the chromosome gets replicated, resulting in an X-shaped structure called a metaphase chromosome. Both the original and the newly copied DNA are now called chromatids. The two "sister" chromatids are joined together at a protein junction called a centromere (forming the X-shaped structure).

Chromosomes are normally visible under a light microscope only when the cell is undergoing mitosis (cell division). Even then, the full chromosome containing both joined sister chromatids becomes visible only during a sequence of mitosis known as metaphase (when chromosomes align together, attached to the mitotic spindle and prepare to divide). This DNA and its associated proteins and macromolecules is collectively known as chromatin, which is further packaged along with its associated molecules into a discrete structure called a nucleosome. Chromatin is present in most cells, with a few exceptions, for example, red blood cells. Occurring only in the nucleus of eukaryotic cells, chromatin composes the vast majority of all DNA, except for a small amount inherited maternally, which is found in mitochondria.

In prokaryotic cells, Chromatin occurs free-floating in cytoplasm, as these cells lack organelles and a defined nucleus. Bacteria also lack histones. The main information-carrying macromolecule is a single piece of coiled double-helix DNA, containing many genes, regulatory elements and other noncoding DNA. The DNA-bound macromolecules are proteins that serve to package the DNA and control its functions. Chromosomes vary widely between different organisms. Some species such as certain bacteria also contain plasmids or other extrachromosomal DNA. These are circular structures in the cytoplasm that contain cellular DNA and play a role in horizontal gene transfer.

Compaction of the duplicated chromosomes during cell division (mitosis or meiosis) results either in a four-arm structure (pictured above) if the centromere is located in the middle of the chromosome or a two-arm structure if the centromere is located near one of the ends. Chromosomal recombination during meiosis and subsequent sexual reproduction plays a significant role in genetic diversity. If these structures are manipulated incorrectly, through processes known as chromosomal instability and translocation, the cell may undergo mitotic catastrophe and die, or it may unexpectedly evade apoptosis leading to the progression of cancer.

In prokaryotes and viruses, the DNA is often densely packed and organized: in the case of archaea, by homologs to eukaryotic histones, and in the case of bacteria, by histone-like proteins. Small circular genomes called plasmids are often found in bacteria and also in mitochondria and chloroplasts, reflecting their bacterial origins.

Some authors, as in this article, use the term chromosome in a wider sense, to refer to the individualized portions of chromatin in cells, either visible or not under light microscopy. However, others use the concept in a narrower sense, to refer to the individualized portions of chromatin during cell division, visible under light microscopy due to high condensation.

History of Discovery

Schleiden, Virchow and Bütschli were among the first scientists who recognized the structures now so familiar to everyone as chromosomes. The term was coined by von Waldeyer-Hartz, referring to the term chromatin, which was introduced by Walther Flemming.

In a series of experiments beginning in the mid-1880s, Theodor Boveri gave the definitive demonstration that chromosomes are the vectors of heredity. His two principles were the *continuity* of chromosomes and the *individuality* of chromosomes. It is the second of these principles that was so original. Wilhelm Roux suggested that each chromosome carries a different genetic load. Boveri was able to test and confirm this hypothesis. Aided by the rediscovery at the start of the 1900s of

Gregor Mendel's earlier work, Boveri was able to point out the connection between the rules of inheritance and the behaviour of the chromosomes. Boveri influenced two generations of American cytologists: Edmund Beecher Wilson, Nettie Stevens, Walter Sutton and Theophilus Painter were all influenced by Boveri (Wilson, Stevens, and Painter actually worked with him).

Walter Sutton (left) and Theodor Boveri (right) independently developed the chromosome theory of inheritance in 1902.

In his famous textbook *The Cell in Development and Heredity*, Wilson linked together the independent work of Boveri and Sutton (both around 1902) by naming the chromosome theory of inheritance the Boveri–Sutton chromosome theory (the names are sometimes reversed). Ernst Mayr remarks that the theory was hotly contested by some famous geneticists: William Bateson, Wilhelm Johannsen, Richard Goldschmidt and T.H. Morgan, all of a rather dogmatic turn of mind. Eventually, complete proof came from chromosome maps in Morgan's own lab.

The number of human chromosomes was published in 1923 by Theophilus Painter. By inspection through the microscope he counted 24 pairs, which would mean 48 chromosomes. His error was copied by others and it was not until 1956 that the true number, 46, was determined by Indonesia-born cytogeneticist Joe Hin Tjio.

Prokaryotes

The prokaryotes – bacteria and archaea – typically have a single circular chromosome, but many variations exist. The chromosomes of most bacteria, which some authors prefer to call genophores, can range in size from only 130,000 base pairs in the endosymbiotic bacteria *Candidatus Hodgkinia cicadicola* and *Candidatus Tremblaya princeps*, to more than 14,000,000 base pairs in the soil-dwelling bacterium *Sorangium cellulosum*. Spirochaetes of the genus *Borrelia* are a notable exception to this arrangement, with bacteria such as *Borrelia burgdorferi*, the cause of Lyme disease, containing a single *linear* chromosome.

Structure in Sequences

Prokaryotic chromosomes have less sequence-based structure than eukaryotes. Bacteria typically have a one-point (the origin of replication) from which replication starts, whereas some archaea contain multiple replication origins. The genes in prokaryotes are often organized in operons, and do not usually contain introns, unlike eukaryotes.

DNA Packaging

Prokaryotes do not possess nuclei. Instead, their DNA is organized into a structure called the nucleoid. The nucleoid is a distinct structure and occupies a defined region of the bacterial cell. This structure is, however, dynamic and is maintained and remodeled by the actions of a range of histone-like proteins, which associate with the bacterial chromosome. In archaea, the DNA in chromosomes is even more organized, with the DNA packaged within structures similar to eukaryotic nucleosomes.

Bacterial chromosomes tend to be tethered to the plasma membrane of the bacteria. In molecular biology application, this allows for its isolation from plasmid DNA by centrifugation of lysed bacteria and pelleting of the membranes (and the attached DNA).

Prokaryotic chromosomes and plasmids are, like eukaryotic DNA, generally supercoiled. The DNA must first be released into its relaxed state for access for transcription, regulation, and replication.

Eukaryotes

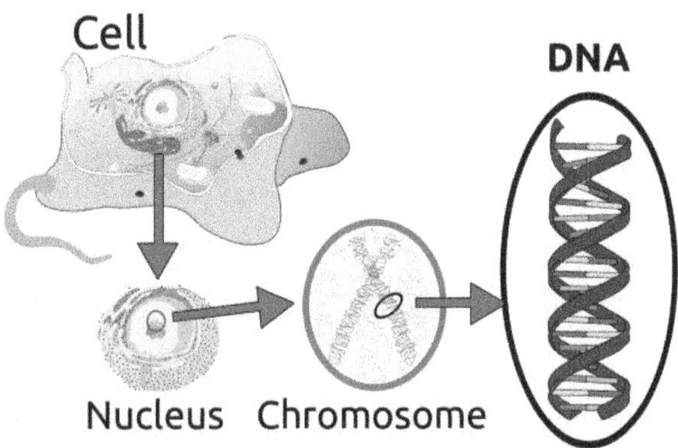

Organization of DNA in a eukaryotic cell.

In eukaryotes, nuclear chromosomes are packaged by proteins into a condensed structure called chromatin. This allows the very long DNA molecules to fit into the cell nucleus. The structure of chromosomes and chromatin varies through the cell cycle. Chromosomes are even more condensed than chromatin and are an essential unit for cellular division. Chromosomes must be replicated, divided, and passed successfully to their daughter cells so as to ensure the genetic diversity and survival of their progeny. Chromosomes may exist as either duplicated or unduplicated. Unduplicated chromosomes are single double helixes, whereas duplicated chromosomes contain two identical copies (called chromatids or sister chromatids) joined by a centromere.

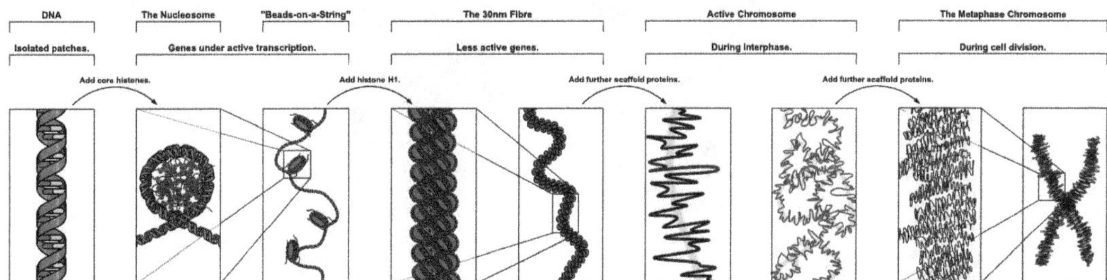

The major structures in DNA compaction: DNA, the nucleosome, the 10 nm "beads-on-a-string" fibre, the 30 nm fibre and the metaphase chromosome.

Eukaryotes (cells with nuclei such as those found in plants, fungi, and animals) possess multiple large linear chromosomes contained in the cell's nucleus. Each chromosome has one centromere, with one or two arms projecting from the centromere, although, under most circumstances, these arms are not visible as such. In addition, most eukaryotes have a small circular mitochondrial genome, and some eukaryotes may have additional small circular or linear cytoplasmic chromosomes.

In the nuclear chromosomes of eukaryotes, the uncondensed DNA exists in a semi-ordered structure, where it is wrapped around histones (structural proteins), forming a composite material called chromatin.

Chromatin

Chromatin is the complex of DNA and protein found in the eukaryotic nucleus, which packages chromosomes. The structure of chromatin varies significantly between different stages of the cell cycle, according to the requirements of the DNA.

Interphase Chromatin

During interphase (the period of the cell cycle where the cell is not dividing), two types of chromatin can be distinguished:

- Euchromatin, which consists of DNA that is active, e.g., being expressed as protein.
- Heterochromatin, which consists of mostly inactive DNA. It seems to serve structural purposes during the chromosomal stages. Heterochromatin can be further distinguished into two types:
    - *Constitutive heterochromatin*, which is never expressed. It is located around the centromere and usually contains repetitive sequences.
    - *Facultative heterochromatin*, which is sometimes expressed.

Metaphase Chromatin and Division

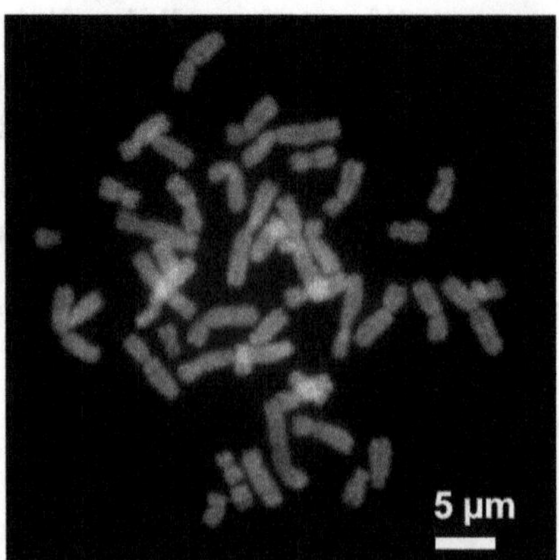

Human chromosomes during metaphase

In the early stages of mitosis or meiosis (cell division), the chromatin double helix become more and more condensed. They cease to function as accessible genetic material (transcription stops) and become a compact transportable form. This compact form makes the individual chromosomes visible, and they form the classic four arm structure, a pair of sister chromatids attached to each other at the centromere. The shorter arms are called *p arms* (from the French *petit*, small) and the longer arms are called *q arms* (*q* follows *p* in the Latin alphabet; q-g "grande"; alternatively it is sometimes said q is short for *queue* meaning tail in French). This is the only natural context in which individual chromosomes are visible with an optical microscope.

Mitotic metaphase chromosomes are best described by a linearly organized longitudinally compressed array of consecutive chromatin loops.

During mitosis, microtubules grow from centrosomes located at opposite ends of the cell and also attach to the centromere at specialized structures called kinetochores, one of which is present on each sister chromatid. A special DNA base sequence in the region of the kinetochores provides, along with special proteins, longer-lasting attachment in this region. The microtubules then pull the chromatids apart toward the centrosomes, so that each daughter cell inherits one set of chromatids. Once the cells have divided, the chromatids are uncoiled and DNA can again be transcribed. In spite of their appearance, chromosomes are structurally highly condensed, which enables these giant DNA structures to be contained within a cell nucleus.

Human Chromosomes

Chromosomes in humans can be divided into two types: autosomes (body chromosome(s)) and allosome (sex chromosome(s)). Certain genetic traits are linked to a person's sex and are passed on through the sex chromosomes. The autosomes contain the rest of the genetic hereditary information. All act in the same way during cell division. Human cells have 23 pairs of chromosomes (22 pairs of autosomes and one pair of sex chromosomes), giving a total of 46 per cell. In addition to these, human cells have many hundreds of copies of the mitochondrial genome. Sequencing of the human genome has provided a great deal of information about each of the chromosomes. Below is a table compiling statistics for the chromosomes, based on the Sanger Institute's human genome information in the Vertebrate Genome Annotation (VEGA) database. Number of genes is an estimate, as it is in part based on gene predictions. Total chromosome length is an estimate as well, based on the estimated size of unsequenced heterochromatin regions.

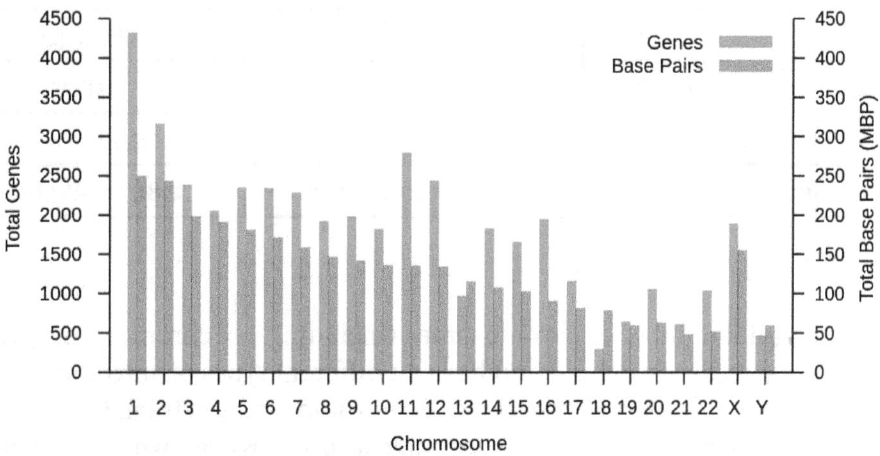

Estimated number of genes and base pairs (in mega base pairs) on each human chromosome

Chromosome	Genes	Total base pairs	% of bases	Sequenced base pairs
1	2000	247,199,719	8.0	224,999,719
2	1300	242,751,149	7.9	237,712,649
3	1000	199,446,827	6.5	194,704,827
4	1000	191,263,063	6.2	187,297,063
5	900	180,837,866	5.9	177,702,766
6	1000	170,896,993	5.5	167,273,993
7	900	158,821,424	5.2	154,952,424
8	700	146,274,826	4.7	142,612,826
9	800	140,442,298	4.6	120,312,298
10	700	135,374,737	4.4	131,624,737
11	1300	134,452,384	4.4	131,130,853
12	1100	132,289,534	4.3	130,303,534
13	300	114,127,980	3.7	95,559,980
14	800	106,360,585	3.5	88,290,585
15	600	100,338,915	3.3	81,341,915
16	800	88,822,254	2.9	78,884,754
17	1200	78,654,742	2.6	77,800,220
18	200	76,117,153	2.5	74,656,155
19	1500	63,806,651	2.1	55,785,651
20	500	62,435,965	2.0	59,505,254
21	200	46,944,323	1.5	34,171,998
22	500	49,528,953	1.6	34,893,953
X (sex chromosome)	800	154,913,754	5.0	151,058,754
Y (sex chromosome)	50	57,741,652	1.9	25,121,652
Total	**21,000**	**3,079,843,747**	**100.0**	**2,857,698,560**

Number in Various Organisms

These tables give the total number of chromosomes (including sex chromosomes) in a cell nucleus. For example, human cells are diploid and have 22 different types of autosome, each present as two copies, and two sex chromosomes. This gives 46 chromosomes in total. Other organisms have more than two copies of their chromosome types, such as bread wheat, which is *hexaploid* and has six copies of seven different chromosome types – 42 chromosomes in total.

Chromosome numbers in some plants	
Plant Species	**#**
Arabidopsis thaliana (diploid)	10
Rye (diploid)	14
Maize (diploid or palaeotetraploid)	20
Einkorn wheat (diploid)	14
Durum wheat (tetraploid)	28
Bread wheat (hexaploid)	42
Cultivated tobacco (tetraploid)	48
Adder's tongue fern (diploid)	approx. 1,200

Chromosome numbers (2n) in some animals			
Species	**#**	**Species**	**#**
Common fruit fly	8	Guinea pig	64
Guppy (*poecilia reticulata*)	46	Garden snail	54
Earthworm (*Octodrilus complanatus*)	36		
Pill millipede (*Arthrosphaera fumosa*)	30		
Tibetan fox	36		
Domestic cat	38	Domestic pig	38
Laboratory mouse	40	Laboratory rat	42
Rabbit (*Oryctolagus cuniculus*)	44	Syrian hamster	44
Hares	48	Human	46
Gorillas, chimpanzees	48	Domestic sheep	54
Elephants	56	Cow	60
Donkey	62	Horse	64
Dog	78		
Hedgehog	90		
Kingfisher	132		
Goldfish	100-104	Silkworm	56

Normal members of a particular eukaryotic species all have the same number of nuclear chromosomes (see the table). Other eukaryotic chromosomes, i.e., mitochondrial and plasmid-like small chromosomes, are much more variable in number, and there may be thousands of copies per cell.

Chromosome numbers in other organisms			
Species	**Large Chromosomes**	**Intermediate Chromosomes**	**Microchromosomes**
Trypanosoma brucei	11	6	~100
Domestic pigeon (*Columba livia domestics*)	18	-	59-63
Chicken	8	2 sex chromosomes	60

Asexually reproducing species have one set of chromosomes that are the same in all body cells. However, asexual species can be either haploid or diploid.

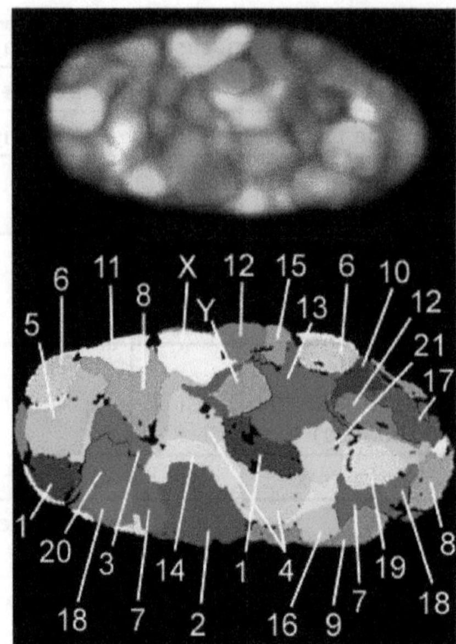

The 23 human chromosome territories during prometaphase in fibroblast cells.

Sexually reproducing species have somatic cells (body cells), which are diploid [2n] having two sets of chromosomes (23 pairs in humans with one set of 23 chromosomes from each parent), one set from the mother and one from the father. Gametes, reproductive cells, are haploid [n]: They have one set of chromosomes. Gametes are produced by meiosis of a diploid germ line cell. During meiosis, the matching chromosomes of father and mother can exchange small parts of themselves (crossover), and thus create new chromosomes that are not inherited solely from either parent. When a male and a female gamete merge (fertilization), a new diploid organism is formed.

Some animal and plant species are polyploid [Xn]: They have more than two sets of homologous chromosomes. Plants important in agriculture such as tobacco or wheat are often polyploid, compared to their ancestral species. Wheat has a haploid number of seven chromosomes, still seen in some cultivars as well as the wild progenitors. The more-common pasta and bread wheats are polyploid, having 28 (tetraploid) and 42 (hexaploid) chromosomes, compared to the 14 (diploid) chromosomes in the wild wheat.

In Prokaryotes

Prokaryote species generally have one copy of each major chromosome, but most cells can easily survive with multiple copies. For example, *Buchnera*, a symbiont of aphids has multiple copies of its chromosome, ranging from 10–400 copies per cell. However, in some large bacteria, such as *Epulopiscium fishelsoni* up to 100,000 copies of the chromosome can be present. Plasmids and plasmid-like small chromosomes are, as in eukaryotes, highly variable in copy number. The number of plasmids in the cell is almost entirely determined by the rate of division of the plasmid – fast division causes high copy number.

Karyotype

In general, the karyotype is the characteristic chromosome complement of a eukaryote species. The preparation and study of karyotypes is part of cytogenetics.

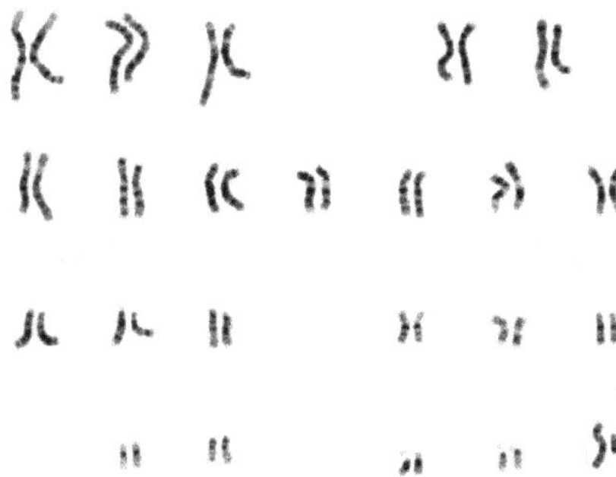

Karyogram of a human male

Although the replication and transcription of DNA is highly standardized in eukaryotes, *the same cannot be said for their karyotypes*, which are often highly variable. There may be variation between species in chromosome number and in detailed organization. In some cases, there is significant variation within species. Often there is:

1. variation between the two sexes
2. variation between the germ-line and soma (between gametes and the rest of the body)
3. variation between members of a population, due to balanced genetic polymorphism
4. geographical variation between races
5. mosaics or otherwise abnormal individuals.

Also, variation in karyotype may occur during development from the fertilized egg.

The technique of determining the karyotype is usually called *karyotyping*. Cells can be locked part-way through division (in metaphase) in vitro (in a reaction vial) with colchicine. These cells are then stained, photographed, and arranged into a *karyogram*, with the set of chromosomes arranged, autosomes in order of length, and sex chromosomes (here X/Y) at the end.

Like many sexually reproducing species, humans have special gonosomes (sex chromosomes, in contrast to autosomes). These are XX in females and XY in males.

Historical Note

Investigation into the human karyotype took many years to settle the most basic question: *How many chromosomes does a normal diploid human cell contain?* In 1912, Hans von Winiwarter reported 47 chromosomes in spermatogonia and 48 in oogonia, concluding an XX/XO sex determination mechanism. Painter in 1922 was not certain whether the diploid number of man is 46 or 48, at first favouring 46. He revised his opinion later from 46 to 48, and he correctly insisted on humans having an XX/XY system.

New techniques were needed to definitively solve the problem:

1. Using cells in culture
2. Arresting mitosis in metaphase by a solution of colchicine

3. Pretreating cells in a hypotonic solution 0.075 M KCl, which swells them and spreads the chromosomes
4. Squashing the preparation on the slide forcing the chromosomes into a single plane
5. Cutting up a photomicrograph and arranging the result into an indisputable karyogram.

It took until 1954 before the human diploid number was confirmed as 46. Considering the techniques of Winiwarter and Painter, their results were quite remarkable. Chimpanzees (the closest living relatives to modern humans) have 48 chromosomes (as well as the other great apes: in humans two chromosomes fused to form chromosome 2).

Aberrations

Chromosomal aberrations are disruptions in the normal chromosomal content of a cell and are a major cause of genetic conditions in humans, such as Down syndrome, although most aberrations have little to no effect. Some chromosome abnormalities do not cause disease in carriers, such as translocations, or chromosomal inversions, although they may lead to a higher chance of bearing a child with a chromosome disorder. Abnormal numbers of chromosomes or chromosome sets, called aneuploidy, may be lethal or may give rise to genetic disorders. Genetic counseling is offered for families that may carry a chromosome rearrangement.

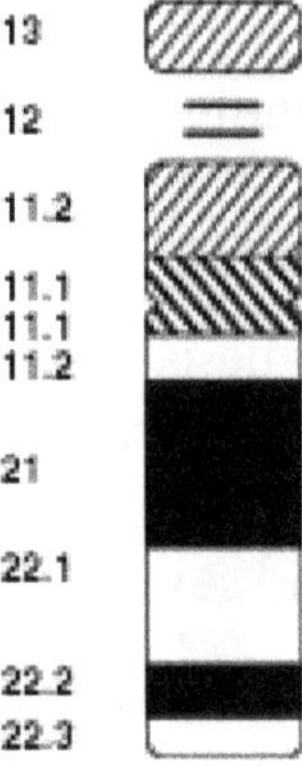

In Down syndrome, there are three copies of chromosome 21

The gain or loss of DNA from chromosomes can lead to a variety of genetic disorders. Human examples include:

- Cri du chat, which is caused by the deletion of part of the short arm of chromosome 5. "Cri du chat" means "cry of the cat" in French; the condition was so-named because affected babies make high-pitched cries that sound like those of a cat. Affected individuals have wide-set eyes, a small head and jaw, moderate to severe mental health problems, and are very short.
- Down syndrome, the most common trisomy, usually caused by an extra copy of chromosome 21 (trisomy 21). Characteristics include decreased muscle tone, stockier build, asymmetrical skull, slanting eyes and mild to moderate developmental disability.
- Edwards syndrome, or trisomy-18, the second most common trisomy. Symptoms include motor retardation, developmental disability and numerous congenital anomalies causing

serious health problems. Ninety percent of those affected die in infancy. They have characteristic clenched hands and overlapping fingers.

- Isodicentric 15, also called idic(15), partial tetrasomy 15q, or inverted duplication 15 (inv dup 15).
- Jacobsen syndrome, which is very rare. It is also called the terminal 11q deletion disorder. Those affected have normal intelligence or mild developmental disability, with poor expressive language skills. Most have a bleeding disorder called Paris-Trousseau syndrome.
- Klinefelter syndrome (XXY). Men with Klinefelter syndrome are usually sterile and tend to be taller and have longer arms and legs than their peers. Boys with the syndrome are often shy and quiet and have a higher incidence of speech delay and dyslexia. Without testosterone treatment, some may develop gynecomastia during puberty.
- Patau Syndrome, also called D-Syndrome or trisomy-13. Symptoms are somewhat similar to those of trisomy-18, without the characteristic folded hand.
- Small supernumerary marker chromosome. This means there is an extra, abnormal chromosome. Features depend on the origin of the extra genetic material. Cat-eye syndrome and isodicentric chromosome 15 syndrome (or Idic15) are both caused by a supernumerary marker chromosome, as is Pallister–Killian syndrome.
- Triple-X syndrome (XXX). XXX girls tend to be tall and thin and have a higher incidence of dyslexia.
- Turner syndrome (X instead of XX or XY). In Turner syndrome, female sexual characteristics are present but underdeveloped. Females with Turner syndrome often have a short stature, low hairline, abnormal eye features and bone development and a “caved-in” appearance to the chest.
- Wolf–Hirschhorn syndrome, which is caused by partial deletion of the short arm of chromosome 4. It is characterized by growth retardation, delayed motor skills development, “Greek Helmet” facial features, and mild to profound mental health problems.
- XYY syndrome. XYY boys are usually taller than their siblings. Like XXY boys and XXX girls, they are more likely to have learning difficulties.

Sperm Aneuploidy

Exposure of males to certain lifestyle, environmental and/or occupational hazards may increase the risk of aneuploid spermatozoa. In particular, risk of aneuploidy is increased by tobacco smoking, and occupational exposure to benzene, insecticides, and perfluorinated compounds. Increased aneuploidy is often associated with increased DNA damage in spermatozoa.

Nucleotide

Nucleotides are organic molecules that serve as the monomers, or subunits, of nucleic acids like DNA (deoxyribonucleic acid) and RNA (ribonucleic acid). The building blocks of nucleic acids, nucleotides are composed of a nitrogenous base, a five-carbon sugar (ribose or deoxyribose), and at least one phosphate group. Thus a nucleoside plus a phosphate group yields a nucleotide.

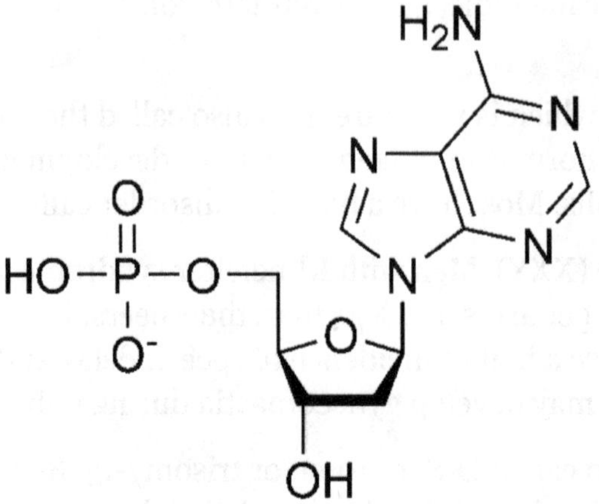

This nucleotide contains the five-carbon sugar deoxyribose, a nitrogenous base called adenine, and one phosphate group. Together, the deoxyribose and adenine make up a nucleoside (specifically, a deoxyribonucleoside) called deoxyadenosine. With the one phosphate group included, whole structure is considered a deoxyribonucleotide (a nucleotide constituent of DNA) with the name deoxyadenosine monophosphate.

Nucleotides also function to carry packets of energy within the cell in the form of the nucleoside triphosphates (ATP, GTP, CTP and UTP), playing a central role in metabolism. In addition, nucleotides participate in cell signaling (cGMP and cAMP), and are incorporated into important cofactors of enzymatic reactions (e.g. coenzyme A, FAD, FMN, NAD, and $NADP^+$).

In experimental biochemistry, nucleotides can be radiolabeled with radionuclides to yield radionucleotides.

Nucleotides are becoming increasingly popular in supplement form.

Structure

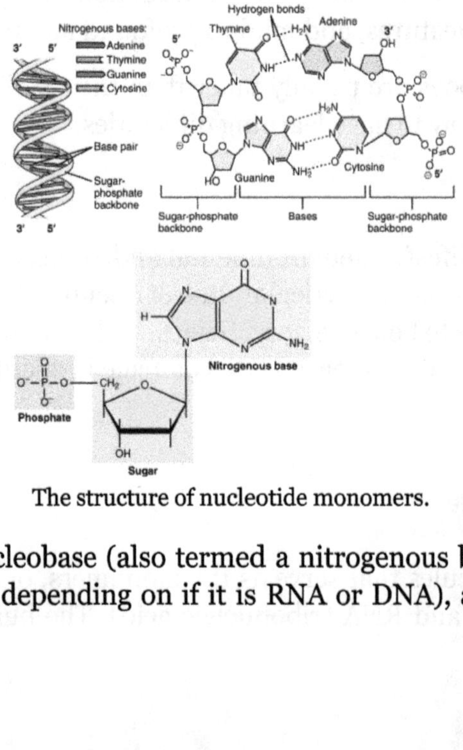

The structure of nucleotide monomers.

A nucleotide is made of a nucleobase (also termed a nitrogenous base), a five-carbon sugar (either ribose or 2-deoxyribose, depending on if it is RNA or DNA), and one or, depending on the definition, more than one phosphate groups. Authoritative chemistry sources such as the ACS Style Guide and IUPAC Gold Book clearly state that the term nucleotide refers only to a molecule containing one phosphate. However, common usage in molecular biology textbooks often extends this definition to include molecules with two or three phosphate groups. Thus, the term "nucleotide" generally refers to a nucleoside monophosphate, but a nucleoside diphosphate or nucleoside triphosphate could be considered a nucleotide as well.

Without the phosphate group, the nucleobase and sugar compose a nucleoside. The phosphate groups form bonds with either the 2, 3, or 5-carbon of the sugar, with the 5-carbon site most common. Cyclic nucleotides form when the phosphate group is bound to two of the sugar's hydroxyl groups. Nucleotides contain either a purine or a pyrimidine base. Ribonucleotides are nucleotides in which the sugar is ribose. Deoxyribonucleotides are nucleotides in which the sugar is deoxyribose.

Nucleic acids are polymeric macromolecules made from nucleotide monomers. In DNA, the purine bases are adenine and guanine, while the pyrimidines are thymine and cytosine. RNA uses uracil in place of thymine. Adenine always pairs with thymine by 2 hydrogen bonds, while guanine pairs with cytosine through 3 hydrogen bonds, in each case because of the unique structures of the bases.

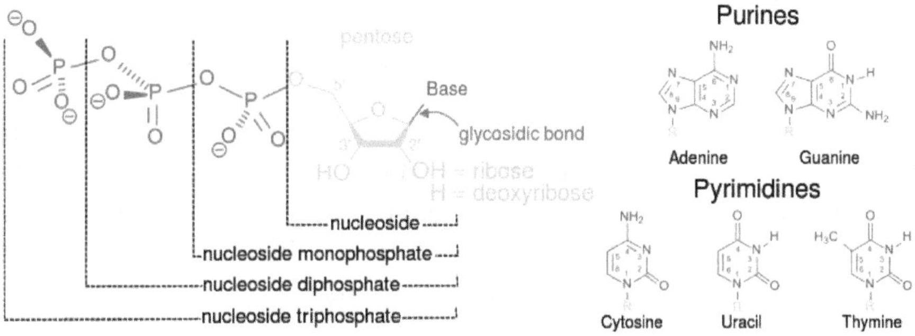

Structural elements of common nucleic acid constituents. Because they contain at least one phosphate group, the compounds marked *nucleoside monophosphate, nucleoside diphosphate* and *nucleoside triphosphate* are all nucleotides (not simply phosphate-lacking nucleosides).

Synthesis

Nucleotides can be synthesized by a variety of means both in vitro and in vivo.

In vivo, nucleotides can be synthesized de novo or recycled through salvage pathways. The components used in de novo nucleotide synthesis are derived from biosynthetic precursors of carbohydrate and amino acid metabolism, and from ammonia and carbon dioxide. The liver is the major organ of de novo synthesis of all four nucleotides. De novo synthesis of pyrimidines and purines follows two different pathways. Pyrimidines are synthesized first from aspartate and carbamoyl-phosphate in the cytoplasm to the common precursor ring structure orotic acid, onto which a phosphorylated ribosyl unit is covalently linked. Purines, however, are first synthesized from the sugar template onto which the ring synthesis occurs. For reference, the syntheses of the purine and pyrimidine nucleotides are carried out by several enzymes in the cytoplasm of the cell, not within a specific organelle. Nucleotides undergo breakdown such that useful parts can be reused in synthesis reactions to create new nucleotides.

In vitro, protecting groups may be used during laboratory production of nucleotides. A purified nucleoside is protected to create a phosphoramidite, which can then be used to obtain analogues not found in nature and/or to synthesize an oligonucleotide.

Pyrimidine Ribonucleotide Synthesis

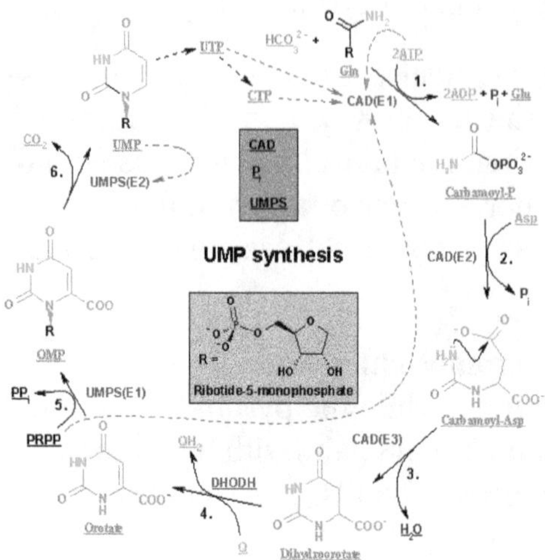

The synthesis of UMP.

The color scheme is as follows: enzymes, coenzymes, substrate names, inorganic molecules

The synthesis of the pyrimidines CTP and UTP occurs in the cytoplasm and starts with the formation of carbamoyl phosphate from glutamine and CO_2. Next, aspartate carbamoyltransferase catalyzes a condensation reaction between aspartate and carbamoyl phosphate to form carbamoyl aspartic acid, which is cyclized into 4,5-dihydroorotic acid by dihydroorotase. The latter is converted to orotate by dihydroorotate oxidase. The net reaction is:

$$(S)\text{-Dihydroorotate} + O_2 \rightarrow \text{Orotate} + H_2O_2$$

Orotate is covalently linked with a phosphorylated ribosyl unit. The covalent linkage between the ribose and pyrimidine occurs at position C_1 of the ribose unit, which contains a pyrophosphate, and N_1 of the pyrimidine ring. Orotate phosphoribosyltransferase (PRPP transferase) catalyzes the net reaction yielding orotidine monophosphate (OMP):

Orotate + 5-Phospho-α-D-ribose 1-diphosphate (PRPP) → Orotidine 5'-phosphate + Pyrophosphate

Orotidine 5'-monophosphate is decarboxylated by orotidine-5'-phosphate decarboxylase to form uridine monophosphate (UMP). PRPP transferase catalyzes both the ribosylation and decarboxylation reactions, forming UMP from orotic acid in the presence of PRPP. It is from UMP that other pyrimidine nucleotides are derived. UMP is phosphorylated by two kinases to uridine triphosphate (UTP) via two sequential reactions with ATP. First the diphosphate form UDP is produced, which in turn is phosphorylated to UTP. Both steps are fueled by ATP hydrolysis:

ATP + UMP → ADP + UDP

UDP + ATP → UTP + ADP

CTP is subsequently formed by amination of UTP by the catalytic activity of CTP synthetase. Glutamine is the NH_3 donor and the reaction is fueled by ATP hydrolysis, too:

$$UTP + Glutamine + ATP + H_2O \rightarrow CTP + ADP + P_i$$

Cytidine monophosphate (CMP) is derived from cytidine triphosphate (CTP) with subsequent loss of two phosphates.

Purine Ribonucleotide Synthesis

The atoms which are used to build the purine nucleotides come from a variety of sources:

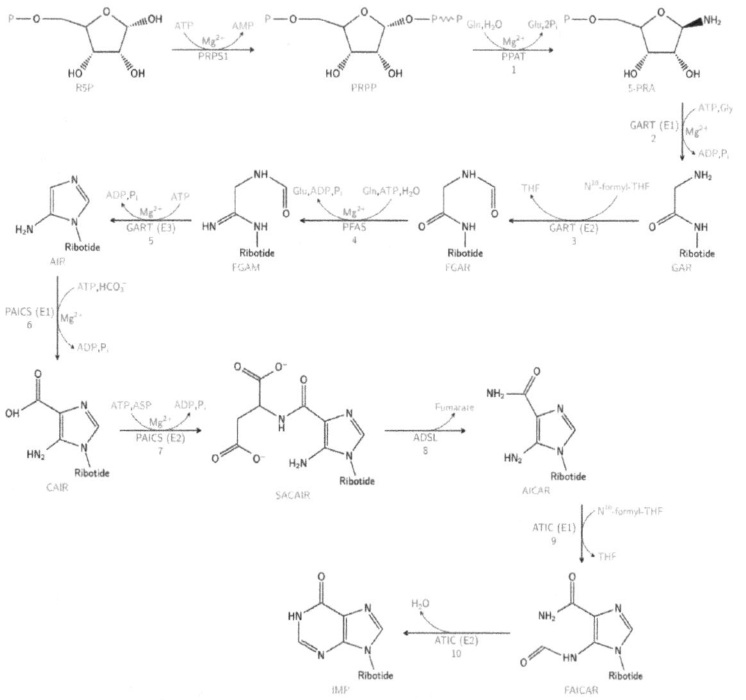

The synthesis of IMP. The color scheme is as follows: enzymes, coenzymes, substrate names, metal ions, inorganic molecules

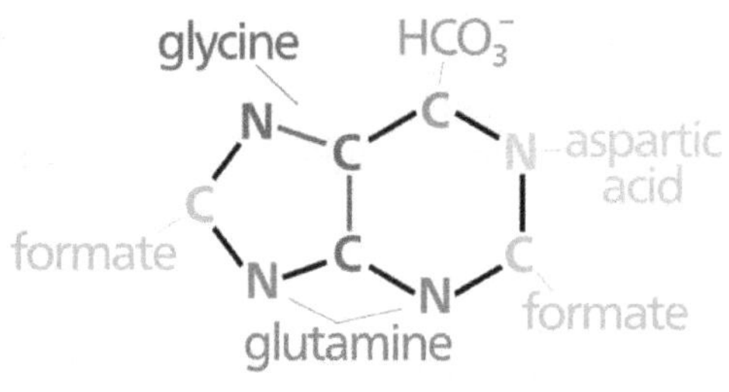

The biosynthetic origins of purine ring atoms

N_1 arises from the amine group of Asp

C_2 and C_8 originate from formate

N_3 and N_9 are contributed by the amide group of Gln

C_4, C_5 and N_7 are derived from Gly

C_6 comes from HCO_3^- (CO_2)

The de novo synthesis of purine nucleotides by which these precursors are incorporated into the purine ring proceeds by a 10-step pathway to the branch-point intermediate IMP, the nucleotide of the base hypoxanthine. AMP and GMP are subsequently synthesized from this intermediate via separate, two-step pathways. Thus, purine moieties are initially formed as part of the ribonucleotides rather than as free bases.

Six enzymes take part in IMP synthesis. Three of them are multifunctional:

- GART (reactions 2, 3, and 5)
- PAICS (reactions 6, and 7)
- ATIC (reactions 9, and 10)

The pathway starts with the formation of PRPP. PRPS1 is the enzyme that activates R5P, which is formed primarily by the pentose phosphate pathway, to PRPP by reacting it with ATP. The reaction is unusual in that a pyrophosphoryl group is directly transferred from ATP to C_1 of R5P and that the product has the α configuration about C1. This reaction is also shared with the pathways for the synthesis of Trp, His, and the pyrimidine nucleotides. Being on a major metabolic crossroad and requiring much energy, this reaction is highly regulated.

In the first reaction unique to purine nucleotide biosynthesis, PPAT catalyzes the displacement of PRPP's pyrophosphate group (PP_i) by an amide nitrogen donated from either glutamine (N), glycine (N&C), aspartate (N), folic acid (C_1), or CO_2. This is the committed step in purine synthesis. The reaction occurs with the inversion of configuration about ribose C_1, thereby forming **β**-5-phosphorybosylamine (5-PRA) and establishing the anomeric form of the future nucleotide.

Next, a glycine is incorporated fueled by ATP hydrolysis and the carboxyl group forms an amine bond to the NH_2 previously introduced. A one-carbon unit from folic acid coenzyme N_{10}-formyl-THF is then added to the amino group of the substituted glycine followed by the closure of the imidazole ring. Next, a second NH_2 group is transferred from a glutamine to the first carbon of the glycine unit. A carboxylation of the second carbon of the glycin unit is concomittantly added. This new carbon is modified by the additional of a third NH_2 unit, this time transferred from an aspartate residue. Finally, a second one-carbon unit from formyl-THF is added to the nitrogen group and the ring covalently closed to form the common purine precursor inosine monophosphate (IMP).

Inosine monophosphate is converted to adenosine monophosphate in two steps. First, GTP hydrolysis fuels the addition of aspartate to IMP by adenylosuccinate synthase, substituting the carbonyl oxygen for a nitrogen and forming the intermediate adenylosuccinate. Fumarate is then cleaved off forming adenosine monophosphate. This step is catalyzed by adenylosuccinate lyase.

Inosine monophosphate is converted to guanosine monophosphate by the oxidation of IMP forming xanthylate, followed by the insertion of an amino group at C_2. NAD^+ is the electron acceptor in the oxidation reaction. The amide group transfer from glutamine is fueled by ATP hydrolysis.

Pyrimidine and Purine Degradation

In humans, pyrimidine rings (C, T, U) can be degraded completely to CO_2 and NH_3 (urea excretion). That having been said, purine rings (G, A) cannot. Instead they are degraded to the metabolically inert uric acid which is then excreted from the body. Uric acid is formed when GMP is split into the base guanine and ribose. Guanine is deaminated to xanthine which in turn is oxidized to uric acid. This last reaction is irreversible. Similarly, uric acid can be formed when AMP is deaminated to IMP from which the ribose unit is removed to form hypoxanthine. Hypoxanthine is oxidized to xanthine and finally to uric acid. Instead of uric acid secretion, guanine and IMP can be used for recycling purposes and nucleic acid synthesis in the presence of PRPP and aspartate (NH_3 donor).

Unnatural Base Pair (UBP)

An unnatural base pair (UBP) is a designed subunit (or nucleobase) of DNA which is created in a laboratory and does not occur in nature. In 2012, a group of American scientists led by Floyd Romesberg, a chemical biologist at the Scripps Research Institute in San Diego, California, published that his team designed an unnatural base pair (UBP). The two new artificial nucleotides or *Unnatural Base Pair* (UBP) were named d5SICS and dNaM. More technically, these artificial nucleotides bearing hydrophobic nucleobases, feature two fused aromatic rings that form a (d5SICS–dNaM) complex or base pair in DNA. In 2014 the same team from the Scripps Research Institute reported that they synthesized a stretch of circular DNA known as a plasmid containing natural T-A and C-G base pairs along with the best-performing UBP Romesberg's laboratory had designed, and inserted it into cells of the common bacterium *E. coli* that successfully replicated the unnatural base pairs through multiple generations. This is the first known example of a living organism passing along an expanded genetic code to subsequent generations. This was in part achieved by the addition of a supportive algal gene that expresses a nucleotide triphosphate transporter which efficiently imports the triphosphates of both d5SICSTP and dNaMTP into *E. coli* bacteria. Then, the natural bacterial replication pathways use them to accurately replicate the plasmid containing d5SICS–dNaM.

The successful incorporation of a third base pair is a significant breakthrough toward the goal of greatly expanding the number of amino acids which can be encoded by DNA, from the existing 20 amino acids to a theoretically possible 172, thereby expanding the potential for living organisms to produce novel proteins. The artificial strings of DNA do not encode for anything yet, but scientists speculate they could be designed to manufacture new proteins which could have industrial or pharmaceutical uses.

Length Unit

Nucleotide (abbreviated "nt") is a common unit of length for single-stranded nucleic acids, similar to how base pair is a unit of length for double-stranded nucleic acids.

Nucleotide Supplements

As a supplement, nucleotides may help the body's immune system and repair the intestines from damage done by excess consumption of alcohol, gluten, casein, etc. The supplement has been studied experimentally.

A study done by the Department of Sports Science at the University of Hull in Hull, UK has shown that nucleotides have significant impact on cortisol levels in saliva. Post exercise, the experimental nucleotide group had lower cortisol levels in their blood than the control or the placebo. Additionally, post supplement values of Immunoglobulin A were significantly higher than either the placebo or the control. The study concluded, "nucleotide supplementation blunts the response of the hormones associated with physiological stress."

Another study conducted in 2013 looked at the impact nucleotide supplementation had on the immune system in athletes. In the study, all athletes were male and were highly skilled in taekwondo. Out of the twenty athletes tested, half received a placebo and half received 480 mg per day of nucleotide supplement. After thirty days, the study concluded that nucleotide supplementation may counteract the impairment of the body's immune function after heavy exercise.

Abbreviation Codes for Degenerate Bases

The IUPAC has designated the symbols for nucleotides. Apart from the five (A, G, C, T/U) bases, often degenerate bases are used especially for designing PCR primers. These nucleotide codes are listed here. Some primer sequences may also include the character "I", which codes for the non-standard nucleotide inosine. Inosine occurs in tRNAs, and will pair with adenine, cytosine, or thymine. This character does not appear in the following table however, because it does not represent a degeneracy. While inosine can serve a similar function as the degeneracy "D", it is an actual nucleotide, rather than a representation of a mix of nucleotides that covers each possible pairing needed.

Symbol	Description	Bases represented				
A	adenine	A				
C	cytosine		C			
G	guanine			G		1
T	thymine				T	
U	uracil				U	
W	weak	A			T	
S	strong		C	G		
M	amino	A	C			
K	keto			G	T	2
R	purine	A		G		
Y	pyrimidine		C		T	
B	not A (**B** comes after A)		C	G	T	
D	not C (**D** comes after C)	A		G	T	
H	not G (**H** comes after G)	A	C		T	3
V	not T (**V** comes after T and U)	A	C	G		
N or -	any base (not a gap)	A	C	G	T	4

Genetic Diversity

Genetic diversity is the total number of genetic characteristics in the genetic makeup of a species. It is distinguished from genetic variability, which describes the tendency of genetic characteristics to vary.

Genetic diversity serves as a way for populations to adapt to changing environments. With more variation, it is more likely that some individuals in a population will possess variations of alleles that are suited for the environment. Those individuals are more likely to survive to produce offspring bearing that allele. The population will continue for more generations because of the success of these individuals.

The academic field of population genetics includes several hypotheses and theories regarding genetic diversity. The neutral theory of evolution proposes that diversity is the result of the accumulation of neutral substitutions. Diversifying selection is the hypothesis that two subpopulations of a species live in different environments that select for different alleles at a particular locus. This may occur, for instance, if a species has a large range relative to the mobility of individuals within it. Frequency-dependent selection is the hypothesis that as alleles become more common, they become more vulnerable. This occurs in host–pathogen interactions, where a high frequency of a defensive allele among the host means that it is more likely that a pathogen will spread if it is able to overcome that allele.

Importance of Genetic Diversity

A 2007 study conducted by the National Science Foundation found that genetic diversity and biodiversity (in terms of species diversity) are dependent upon each other—that diversity within a species is necessary to maintain diversity among species, and vice versa. According to the lead researcher in the study, Dr. Richard Lankau, "If any one type is removed from the system, the cycle can break down, and the community becomes dominated by a single species." Genotypic and phenotypic diversity have been found in all species at the protein, DNA, and organismal levels; in nature, this diversity is nonrandom, heavily structured, and correlated with environmental variation and stress.

The interdependence between genetic and species diversity is delicate. Changes in species diversity lead to changes in the environment, leading to adaptation of the remaining species. Changes in genetic diversity, such as in loss of species, leads to a loss of biological diversity. Loss of genetic diversity in domestic animal populations has also been studied and attributed to the extension of markets and economic globalization.

Survival and Adaptation

Genetic diversity plays an important role in the survival and adaptability of a species. When a population's habitat changes, the population may have to adapt to survive; the ability of the population to adapt to the changing environment will determine their ability to cope with an environmental challenge. Variation in the population's gene pool provides variable traits among the individuals of that population. These variable traits can be selected for, via natural selection, ultimately leading to an adaptive change in the population, allowing it to survive in the changed environment. If a population of a species has a very diverse gene pool then there will be more variety in the traits of individuals of that population and consequently more traits for natural selection to act upon to select the fittest individuals to survive.

Genetic diversity is essential for a species to evolve. With very little gene variation within the species, healthy reproduction becomes increasingly difficult, and offspring are more likely to have problems resulting from inbreeding. The vulnerability of a population to certain types of diseases can also increase with reduction in genetic diversity. Concerns about genetic diversity are especially important with large mammals due to their small population size and high levels of human-caused population effects.

Agricultural Relevance

When humans initially started farming, they used selective breeding to pass on desirable traits of the crops while omitting the undesirable ones. Selective breeding leads to monocultures: entire farms of nearly genetically identical plants. Little to no genetic diversity makes crops extremely susceptible to widespread disease; bacteria morph and change constantly and when a disease-causing bacterium changes to attack a specific genetic variation, it can easily wipe out vast quantities of the species. If the genetic variation that the bacterium is best at attacking happens to be that which humans have selectively bred to use for harvest, the entire crop will be wiped out.

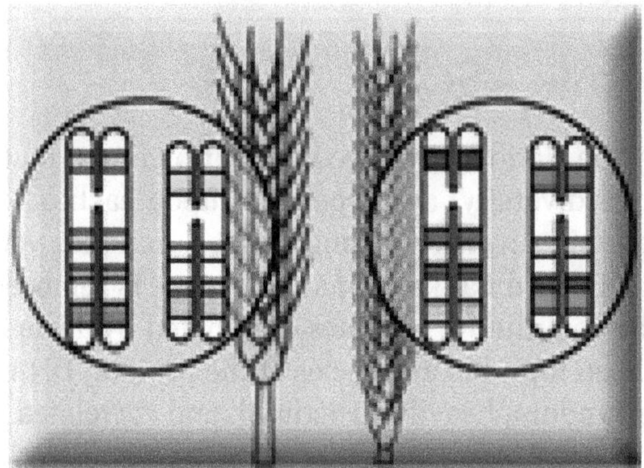

A very similar occurrence is the cause of the infamous Potato Famine in Ireland. Since new potato plants do not come as a result of reproduction, but rather from pieces of the parent plant, no genetic diversity is developed, and the entire crop is essentially a clone of one potato, it is especially susceptible to an epidemic. In the 1840s, much of Ireland's population depended on potatoes for food. They planted namely the "lumper" variety of potato, which was susceptible to a rot-causing oomycete called *Phytophthora infestans*. This oomycete destroyed the vast majority of the potato crop, and left one million people to starve to death.

Livestock Biodiversity

The genetic diversity of livestock species allows animal production to be practiced in a range of environments and with a range of different objectives. It provides the raw material for selective breeding programmes and allows livestock populations to adapt as environmental conditions change.

Livestock biodiversity can be lost as a result of breed extinctions and other forms of genetic erosion. As of June 2014, among the 8 774 breeds recorded in the Domestic Animal Diversity Information System (DAD-IS), operated by the Food and Agriculture Organization of the United Nations (FAO), 17 percent were classified as being at risk of extinction and 7 percent already extinct. There is now a Global Plan of Action for Animal Genetic Resources that was developed under the auspices of the Commission on Genetic Resources for Food and Agriculture in 2007, that provides a framework and guidelines for the management of animal genetic resources.

Awareness of the importance of maintaining animal genetic resources has increased over time. FAO has published two reports on the state of the world's animal genetic resources for food and agriculture, which cover detailed analyses of our global livestock diversity and ability to manage and conserve them.

Coping with Low Genetic Diversity

The natural world has several ways of preserving or increasing genetic diversity. Among oceanic plankton, viruses aid in the genetic shifting process. Ocean viruses, which infect the plankton, carry genes of other organisms in addition to their own. When a virus containing the genes of one cell infects another, the genetic makeup of the latter changes. This constant shift of genetic makeup helps to maintain a healthy population of plankton despite complex and unpredictable environmental changes.

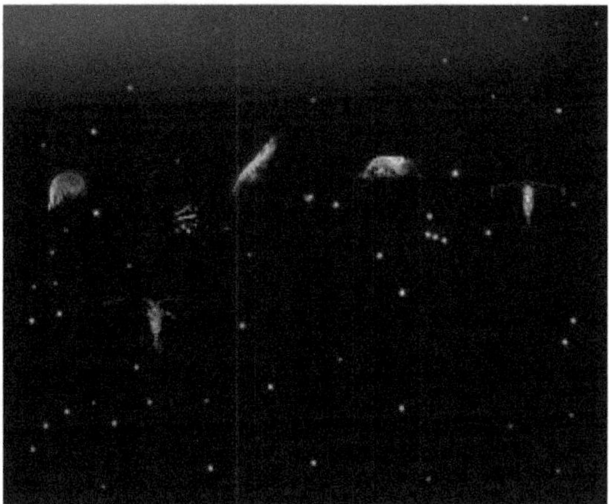

Photomontage of planktonic organisms.

A Tanzanian cheetah.

Cheetahs are a threatened species. Low genetic diversity and resulting poor sperm quality has made breeding and survivorship difficult for cheetahs. Moreover, only about 5% of cheetahs survive to adulthood. However, it has been recently discovered that female cheetahs can mate with more than one male per litter of cubs. They undergo induced ovulation, which means that a new egg is produced every time a female mates. By mating with multiple males, the mother increases the genetic diversity within a single litter of cubs.

Measures of Genetic Diversity

Genetic Diversity of a population can be assessed by some simple measures.

- Gene Diversity is the proportion of polymorphic loci across the genome.
- Heterozygosity is the fraction of individuals in a population that are heterozygous for a particular locus.
- Alleles per locus is also used to demonstrate variability.
- Nucleotide diversity is the extent of nucleotide polymorphisms within a population, and is commonly measured through molecular markers such as micro- and minisatellite sequences, mitochondrial DNA, and single-nucleotide polymorphisms (SNPs).

Other Measures of Diversity

Alternatively, other types of diversity may be assessed for organisms:

- species diversity
- ecological diversity
- morphological diversity
- degeneracy

There are broad correlations between different types of diversity. For example, there is a close link between vertebrate taxonomic and ecological diversity.

Genetic Drift

Genetic drift (also known as allelic drift or the Sewall Wright effect after biologist Sewall Wright) is the change in the frequency of a gene variant (allele) in a population due to random sampling of organisms. The alleles in the offspring are a sample of those in the parents, and chance has a role in determining whether a given individual survives and reproduces. A population's allele frequency is the fraction of the copies of one gene that share a particular form. Genetic drift may cause gene variants to disappear completely and thereby reduce genetic variation.

When there are few copies of an allele, the effect of genetic drift is larger, and when there are many copies the effect is smaller. In the early 20th century, vigorous debates occurred over the relative importance of natural selection versus neutral processes, including genetic drift. Ronald Fisher, who explained natural selection using Mendelian genetics, held the view that genetic drift plays at the most a minor role in evolution, and this remained the dominant view for several decades. In

1968, population geneticist Motoo Kimura rekindled the debate with his neutral theory of molecular evolution, which claims that most instances where a genetic change spreads across a population (although not necessarily changes in phenotypes) are caused by genetic drift acting on neutral mutations.

Biologist and statistician Ronald Fisher

Analogy with Marbles in a Jar

The process of genetic drift can be illustrated using 20 marbles in a jar to represent 20 organisms in a population. Consider this jar of marbles as the starting population. Half of the marbles in the jar are red and half blue, and both colours correspond to two different alleles of one gene in the population. In each new generation the organisms reproduce at random. To represent this reproduction, randomly select a marble from the original jar and deposit a new marble with the same colour as its "offspring" into a new jar. (The selected marble remains in the original jar.) Repeat this process until there are 20 new marbles in the second jar. The second jar then contains a second generation of "offspring," consisting of 20 marbles of various colours. Unless the second jar contains exactly 10 red marbles and 10 blue marbles, a random shift occurred in the allele frequencies.

Repeat this process a number of times, randomly reproducing each generation of marbles to form the next. The numbers of red and blue marbles picked each generation fluctuates; sometimes more red and sometimes more blue. This fluctuation is analogous to genetic drift – a change in the population's allele frequency resulting from a random variation in the distribution of alleles from one generation to the next.

It is even possible that in any one generation no marbles of a particular colour are chosen, meaning they have no offspring. In this example, if no red marbles are selected, the jar representing the new generation contains only blue offspring. If this happens, the red allele has been lost permanently in the population, while the remaining blue allele has become fixed: all future generations are entirely blue. In small populations, fixation can occur in just a few generations.

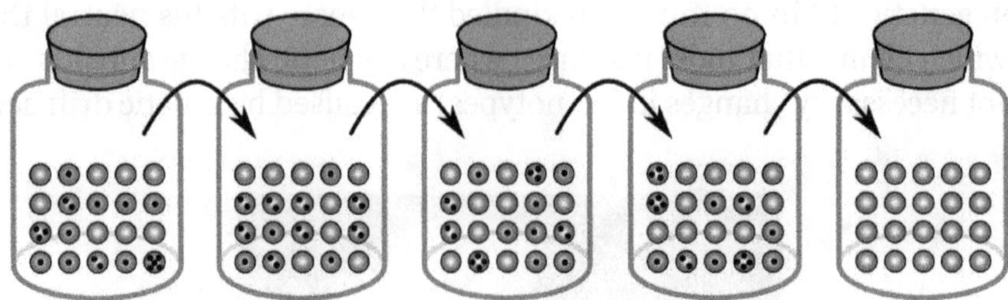

In this simulation, there is fixation in the blue "allele" within five generations.

Probability and Allele Frequency

The mechanisms of genetic drift can be illustrated with a simplified example. Consider a very large colony of bacteria isolated in a drop of solution. The bacteria are genetically identical except for a single gene with two alleles labeled A and B. A and B are neutral alleles meaning that they do not affect the bacteria's ability to survive and reproduce; all bacteria in this colony are equally likely to survive and reproduce. Suppose that half the bacteria have allele A and the other half have allele B. Thus A and B each have allele frequency 1/2.

The drop of solution then shrinks until it has only enough food to sustain four bacteria. All other bacteria die without reproducing. Among the four who survive, there are sixteen possible combinations for the A and B alleles:

**(A-A-A-A), (B-A-A-A), (A-B-A-A), (B-B-A-A),
(A-A-B-A), (B-A-B-A), (A-B-B-A), (B-B-B-A),
(A-A-A-B), (B-A-A-B), (A-B-A-B), (B-B-A-B),
(A-A-B-B), (B-A-B-B), (A-B-B-B), (B-B-B-B).**

Since all bacteria in the original solution are equally likely to survive when the solution shrinks, the four survivors are a random sample from the original colony. The probability that each of the four survivors has a given allele is 1/2, and so the probability that any particular allele combination occurs when the solution shrinks is

$$\frac{1}{2}\cdot\frac{1}{2}\cdot\frac{1}{2}\cdot\frac{1}{2}=\frac{1}{16}.$$

(The original population size is so large that the sampling effectively happens without replacement). In other words, each of the sixteen possible allele combinations is equally likely to occur, with probability 1/16.

Counting the combinations with the same number of A and B, we get the following table.

A	B	Combinations	Probability
4	0	1	1/16
3	1	4	4/16
2	2	6	6/16
1	3	4	4/16
0	4	1	1/16

As shown in the table, the total number of possible combinations to have an equal (conserved) number of A and B alleles is six, and its probability is 6/16. The total number of possible alternative combinations is ten, and the probability of unequal number of A and B alleles is 10/16. Thus, although the original colony began with an equal number of A and B alleles, chances are that the number of alleles in the remaining population of four members will not be equal. In the latter case, genetic drift has occurred because the population's allele frequencies have changed due to random sampling. In this example the population contracted to just four random survivors, a phenomenon known as population bottleneck.

The probabilities for the number of copies of allele A (or B) that survive (given in the last column of the above table) can be calculated directly from the binomial distribution where the "success" probability (probability of a given allele being present) is 1/2 (i.e., the probability that there are k copies of A (or B) alleles in the combination) is given by

$$\binom{n}{k}\left(\frac{1}{2}\right)^{k}\left(1-\frac{1}{2}\right)^{n-k}=\binom{n}{k}\left(\frac{1}{2}\right)^{n}$$

where $n=4$ is the number of surviving bacteria.

Mathematical Models of Genetic Drift

Mathematical models of genetic drift can be designed using either branching processes or a diffusion equation describing changes in allele frequency in an idealised population.

Wright–Fisher Model

Think of a gene with two alleles, A or B. In diploid populations consisting of N individuals there are $2N$ copies of each gene. An individual can have two copies of the same allele or two different alleles. We can call the frequency of one allele p and the frequency of the other q. The Wright–Fisher model (named after Sewall Wright and Ronald Fisher) assumes that generations do not overlap (for example, annual plants have exactly one generation per year) and that each copy of the gene found in the new generation is drawn independently at random from all copies of the gene in the old generation. The formula to calculate the probability of obtaining k copies of an allele that had frequency p in the last generation is then

$$\frac{(2N)!}{k!(2N-k)!}p^{k}q^{2N-k}$$

where the symbol "!" signifies the factorial function. This expression can also be formulated using the binomial coefficient,

$$\binom{2N}{k}p^{k}q^{2N-k}$$

Moran Model

The Moran model assumes overlapping generations. At each time step, one individual is chosen to reproduce and one individual is chosen to die. So in each timestep, the number of copies of a given allele can go up by one, go down by one, or can stay the same. This means that the transition matrix is tridiagonal, which means that mathematical solutions are easier for the Moran model than for the Wright–Fisher model. On the other hand, computer simulations are usually easier to perform using the Wright–Fisher model, because fewer time steps need to be calculated. In the Moran model, it takes N timesteps to get through one generation, where N is the effective population size. In the Wright–Fisher model, it takes just one.

In practice, the Moran model and Wright–Fisher model give qualitatively similar results, but genetic drift runs twice as fast in the Moran model.

Other Models of Drift

If the variance in the number of offspring is much greater than that given by the binomial distribution assumed by the Wright–Fisher model, then given the same overall speed of genetic drift (the variance effective population size), genetic drift is a less powerful force compared to selection. Even for the same variance, if higher moments of the offspring number distribution exceed those of the binomial distribution then again the force of genetic drift is substantially weakened.

Random Effects Other than Sampling Error

Random changes in allele frequencies can also be caused by effects other than sampling error, for example random changes in selection pressure.

One important alternative source of stochasticity, perhaps more important than genetic drift, is genetic draft. Genetic draft is the effect on a locus by selection on linked loci. The mathematical properties of genetic draft are different from those of genetic drift. The direction of the random change in allele frequency is autocorrelated across generations.

Drift and Fixation

The Hardy–Weinberg principle states that within sufficiently large populations, the allele frequencies remain constant from one generation to the next unless the equilibrium is disturbed by migration, genetic mutations, or selection.

However, in finite populations, no new alleles are gained from the random sampling of alleles passed to the next generation, but the sampling can cause an existing allele to disappear. Because random sampling can remove, but not replace, an allele, and because random declines or increases in allele frequency influence expected allele distributions for the next generation, genetic drift drives a population towards genetic uniformity over time. When an allele reaches a frequency of 1 (100%) it is said to be "fixed" in the population and when an allele reaches a frequency of 0 (0%) it is lost. Smaller populations achieve fixation faster, whereas in the limit of an infinite population, fixation is not achieved. Once an allele becomes fixed, genetic drift comes to a halt, and the allele frequency cannot change unless a new allele is introduced in the population via mutation or gene flow. Thus even while genetic drift is a random, directionless process, it acts to eliminate genetic variation over time.

Rate of Allele Frequency Change Due to Drift

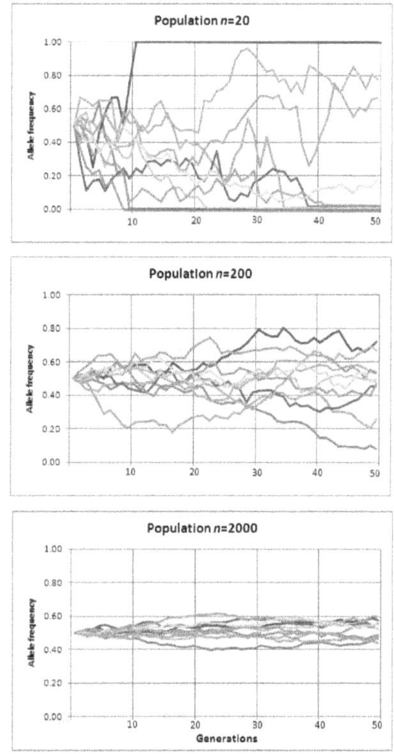

Ten simulations of random genetic drift of a single given allele with an initial frequency distribution 0.5 measured over the course of 50 generations, repeated in three reproductively synchronous populations of different sizes. In these simulations, alleles drift to loss or fixation (frequency of 0.0 or 1.0) only in the smallest population.

Assuming genetic drift is the only evolutionary force acting on an allele, after *t* generations in many replicated populations, starting with allele frequencies of *p* and *q*, the variance in allele frequency across those populations is

$$V_t \approx pq\left(1-\exp\left(-\frac{t}{2N_e}\right)\right)$$

Time to Fixation or Loss

Assuming genetic drift is the only evolutionary force acting on an allele, at any given time the probability that an allele will eventually become fixed in the population is simply its frequency in the population at that time. For example, if the frequency *p* for allele A is 75% and the frequency *q* for allele B is 25%, then given unlimited time the probability A will ultimately become fixed in the population is 75% and the probability that B will become fixed is 25%.

The expected number of generations for fixation to occur is proportional to the population size, such that fixation is predicted to occur much more rapidly in smaller populations. Normally the effective population size, which is smaller than the total population, is used to determine these probabilities. The effective population (N_e) takes into account factors such as the level of inbreeding, the stage of the lifecycle in which the population is the smallest, and the fact that some neutral genes are genetically linked to others that are under selection. The effective population size may not be the same for every gene in the same population.

One forward-looking formula used for approximating the expected time before a neutral allele becomes fixed through genetic drift, according to the Wright–Fisher model, is

$$\bar{T}_{\text{fixed}} = \frac{-4N_e(1-p)\ln(1-p)}{p}$$

where T is the number of generations, N_e is the effective population size, and p is the initial frequency for the given allele. The result is the number of generations expected to pass before fixation occurs for a given allele in a population with given size (N_e) and allele frequency (p).

The expected time for the neutral allele to be lost through genetic drift can be calculated as

$$\bar{T}_{\text{lost}} = \frac{-4N_e p}{1-p}\ln p.$$

When a mutation appears only once in a population large enough for the initial frequency to be negligible, the formulas can be simplified to

$$\bar{T}_{\text{fixed}} = 4N_e$$

for average number of generations expected before fixation of a neutral mutation, and

$$\bar{T}_{\text{lost}} = 2\left(\frac{N_e}{N}\right)\ln(2N)$$

for the average number of generations expected before the loss of a neutral mutation.

Time to Loss with Both Drift and Mutation

The formulae above apply to an allele that is already present in a population, and which is subject to neither mutation nor natural selection. If an allele is lost by mutation much more often than it is gained by mutation, then mutation, as well as drift, may influence the time to loss. If the allele prone to mutational loss begins as fixed in the population, and is lost by mutation at rate m per replication, then the expected time in generations until its loss in a haploid population is given by

$$\bar{T}_{\text{lost}} \approx \begin{cases} \dfrac{1}{m}, & \text{if } mN_e \ll 1 \\ \dfrac{\ln(mN_e)+\gamma}{m} & \text{if } mN_e \gg 1 \end{cases}$$

where γ is equal to Euler's constant. The first approximation represents the waiting time until the first mutant destined for loss, with loss then occurring relatively rapidly by genetic drift, taking time $N_e << 1/m$. The second approximation represents the time needed for deterministic loss by mutation accumulation. In both cases, the time to fixation is dominated by mutation via the term $1/m$, and is less affected by the effective population size.

Genetic Drift Versus Natural Selection

In natural populations, genetic drift and natural selection do not act in isolation; both forces are always at play, together with mutation and migration. Neutral evolution is the product of both mutation and drift, not of drift alone. Similarly, even when selection overwhelms genetic drift, it can only act on variation that mutation provides.

While natural selection has a direction, guiding evolution towards heritable adaptations to the current environment, genetic drift has no direction and is guided only by the mathematics of chance. As a result, drift acts upon the genotypic frequencies within a population without regard to their phenotypic effects. In contrast, selection favors the spread of alleles whose phenotypic effects increase survival and/or reproduction of their carriers, lowers the frequencies of alleles that cause unfavorable traits, and ignores those that are neutral.

The law of large numbers predicts that when the absolute number of copies of the allele is small (e.g., in small populations), the magnitude of drift on allele frequencies per generation is larger. The magnitude of drift is large enough to overwhelm selection at any allele frequency when the selection coefficient is less than 1 divided by the effective population size. Non-adaptive evolution resulting from the product of mutation and genetic drift is therefore considered to be a consequential mechanism of evolutionary change primarily within small, isolated populations. The mathematics of genetic drift depend on the effective population size, but it is not clear how this is related to the actual number of individuals in a population. Genetic linkage to other genes that are under selection can reduce the effective population size experienced by a neutral allele. With a higher recombination rate, linkage decreases and with it this local effect on effective population size. This effect is visible in molecular data as a correlation between local recombination rate and genetic diversity, and negative correlation between gene density and diversity at noncoding DNA regions. Stochasticity associated with linkage to other genes that are under selection is not the same as sampling error, and is sometimes known as genetic draft in order to distinguish it from genetic drift.

When the allele frequency is very small, drift can also overpower selection even in large populations. For example, while disadvantageous mutations are usually eliminated quickly in large populations, new advantageous mutations are almost as vulnerable to loss through genetic drift as are neutral mutations. Not until the allele frequency for the advantageous mutation reaches a certain threshold will genetic drift have no effect.

Population Bottleneck

A population bottleneck is when a population contracts to a significantly smaller size over a short period of time due to some random environmental event. In a true population bottleneck, the odds for survival of any member of the population are purely random, and are not improved by any particular inherent genetic advantage. The bottleneck can result in radical changes in allele frequencies, completely independent of selection.

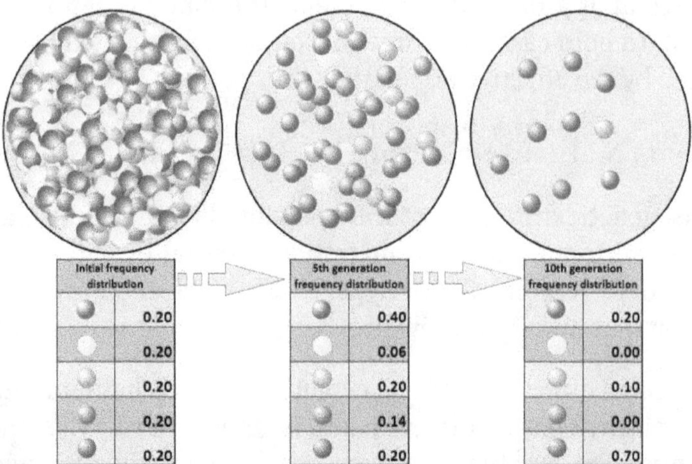

Changes in a population's allele frequency following a population bottleneck: the rapid and radical decline in population size has reduced the population's genetic variation.

The impact of a population bottleneck can be sustained, even when the bottleneck is caused by a one-time event such as a natural catastrophe. An interesting example of a bottleneck causing unusual genetic distribution is the relatively high proportion of individuals with total rod cell color blindness (achromatopsia) on Pingelap atoll in Micronesia. After a bottleneck, inbreeding increases. This increases the damage done by recessive deleterious mutations, in a process known as inbreeding depression. The worst of these mutations are selected against, leading to the loss of other alleles that are genetically linked to them, in a process of background selection. For recessive harmful mutations, this selection can be enhanced as a consequence of the bottleneck, due to genetic purging.This leads to a further loss of genetic diversity. In addition, a sustained reduction in population size increases the likelihood of further allele fluctuations from drift in generations to come.

A population's genetic variation can be greatly reduced by a bottleneck, and even beneficial adaptations may be permanently eliminated. The loss of variation leaves the surviving population vulnerable to any new selection pressures such as disease, climate change or shift in the available food source, because adapting in response to environmental changes requires sufficient genetic variation in the population for natural selection to take place.

There have been many known cases of population bottleneck in the recent past. Prior to the arrival of Europeans, North American prairies were habitat for millions of greater prairie chickens. In Illinois alone, their numbers plummeted from about 100 million birds in 1900 to about 50 birds in the 1990s. The declines in population resulted from hunting and habitat destruction, but the random consequence has been a loss of most of the species' genetic diversity. DNA analysis comparing birds from the mid century to birds in the 1990s documents a steep decline in the genetic variation in just in the latter few decades. Currently the greater prairie chicken is experiencing low reproductive success. However, bottleneck and genetic drift can lead to a genetic loss that increases fitness as seen in Ehrlichia.

Over-hunting also caused a severe population bottleneck in the northern elephant seal in the 19th century. Their resulting decline in genetic variation can be deduced by comparing it to that of the southern elephant seal, which were not so aggressively hunted.

Founder Effect

The founder effect is a special case of a population bottleneck, occurring when a small group in a population splinters off from the original population and forms a new one. The random sample of alleles in the just formed new colony is expected to grossly misrepresent the original population in at least some respects. It is even possible that the number of alleles for some genes in the original population is larger than the number of gene copies in the founders, making complete representation impossible. When a newly formed colony is small, its founders can strongly affect the population's genetic make-up far into the future.

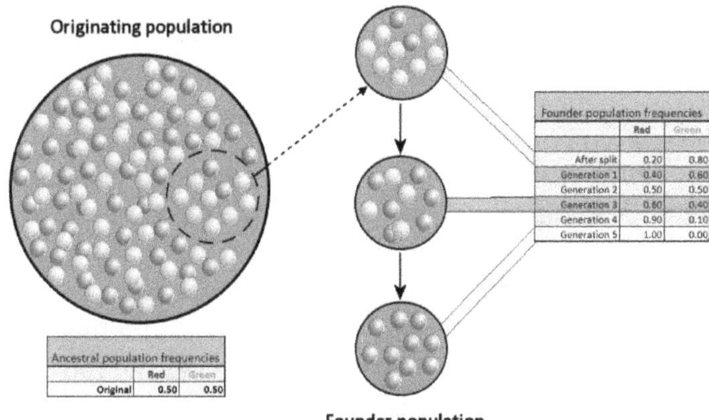

When very few members of a population migrate to form a separate new population, the founder effect occurs. For a period after the foundation, the small population experiences intensive drift. In the figure this results in fixation of the red allele.

A well-documented example is found in the Amish migration to Pennsylvania in 1744. Two members of the new colony shared the recessive allele for Ellis–van Creveld syndrome. Members of the colony and their descendants tend to be religious isolates and remain relatively insular. As a result of many generations of inbreeding, Ellis-van Creveld syndrome is now much more prevalent among the Amish than in the general population.

The difference in gene frequencies between the original population and colony may also trigger the two groups to diverge significantly over the course of many generations. As the difference, or genetic distance, increases, the two separated populations may become distinct, both genetically and phenetically, although not only genetic drift but also natural selection, gene flow, and mutation contribute to this divergence. This potential for relatively rapid changes in the colony's gene frequency led most scientists to consider the founder effect (and by extension, genetic drift) a significant driving force in the evolution of new species. Sewall Wright was the first to attach this significance to random drift and small, newly isolated populations with his shifting balance theory of speciation. Following after Wright, Ernst Mayr created many persuasive models to show that the decline in genetic variation and small population size following the founder effect were critically important for new species to develop. However, there is much less support for this view today since the hypothesis has been tested repeatedly through experimental research and the results have been equivocal at best.

Founder effect was first well-investigated in the USSR by Soviet scientists Lisovskiy V.V., Kuznetsov M.A. and Nikolay Dubinin.

History of the Concept

The concept of genetic drift was first introduced by one of the founders in the field of population genetics, Sewall Wright. His first use of the term "drift" was in 1929, though at the time he was using it in the sense of a directed process of change, or natural selection. Random drift by means of sampling error came to be known as the "Sewall–Wright effect," though he was never entirely comfortable to see his name given to it. Wright referred to all changes in allele frequency as either "steady drift" (e.g., selection) or "random drift" (e.g., sampling error). "Drift" came to be adopted as a technical term in the stochastic sense exclusively. Today it is usually defined still more narrowly, in terms of sampling error, although this narrow definition is not universal. Wright wrote that the "restriction of "random drift" or even "drift" to only one component, the effects of accidents of sampling, tends to lead to confusion." Sewall Wright considered the process of random genetic drift by means of sampling error equivalent to that by means of inbreeding, but later work has shown them to be distinct.

In the early days of the modern evolutionary synthesis, scientists were just beginning to blend the new science of population genetics with Charles Darwin's theory of natural selection. Working within this new framework, Wright focused on the effects of inbreeding on small relatively isolated populations. He introduced the concept of an adaptive landscape in which phenomena such as cross breeding and genetic drift in small populations could push them away from adaptive peaks, which in turn allow natural selection to push them towards new adaptive peaks. Wright thought smaller populations were more suited for natural selection because "inbreeding was sufficiently intense to create new interaction systems through random drift but not intense enough to cause random nonadaptive fixation of genes."

Wright's views on the role of genetic drift in the evolutionary scheme were controversial almost from the very beginning. One of the most vociferous and influential critics was colleague Ronald Fisher. Fisher conceded genetic drift played some role in evolution, but an insignificant one. Fisher has been accused of misunderstanding Wright's views because in his criticisms Fisher seemed to argue Wright had rejected selection almost entirely. To Fisher, viewing the process of evolution as a long, steady, adaptive progression was the only way to explain the ever-increasing complexity from simpler forms. But the debates have continued between the "gradualists" and those who lean more toward the Wright model of evolution where selection and drift together play an important role.

In 1968, Motoo Kimura rekindled the debate with his neutral theory of molecular evolution, which claims that most of the genetic changes are caused by genetic drift acting on neutral mutations.

The role of genetic drift by means of sampling error in evolution has been criticized by John H. Gillespie and William B. Provine, who argue that selection on linked sites is a more important stochastic force.

Mutation

In biology, a mutation is the permanent alteration of the nucleotide sequence of the genome of an organism, virus, or extrachromosomal DNA or other genetic elements. Mutations result from errors during DNA replication or other types of damage to DNA, which then may undergo error-prone repair (especially microhomology-mediated end joining), or cause an error during other forms of repair, or else may cause an error during replication (translesion synthesis). Mutations may also result from insertion or deletion of segments of DNA due to mobile genetic elements. Mutations may or may not produce discernible changes in the observable characteristics (phenotype) of an organism. Mutations play a part in both normal and abnormal biological processes including: evolution, cancer, and the development of the immune system, including junctional diversity.

The genomes of RNA viruses are based on RNA rather than DNA. The RNA viral genome can be double stranded (as in DNA) or single stranded. In some of these viruses (such as the single stranded human immunodeficiency virus) replication occurs quickly and there are no mechanisms to check the genome for accuracy. This error-prone process often results in mutations.

Mutation can result in many different types of change in sequences. Mutations in genes can either have no effect, alter the product of a gene, or prevent the gene from functioning properly or completely. Mutations can also occur in nongenic regions. One study on genetic variations between different species of *Drosophila* suggests that, if a mutation changes a protein produced by a gene, the result is likely to be harmful, with an estimated 70 percent of amino acid polymorphisms that have damaging effects, and the remainder being either neutral or marginally beneficial. Due to the damaging effects that mutations can have on genes, organisms have mechanisms such as DNA repair to prevent or correct mutations by reverting the mutated sequence back to its original state.

Description

Mutations can involve the duplication of large sections of DNA, usually through genetic recombination. These duplications are a major source of raw material for evolving new genes, with tens to hundreds of genes duplicated in animal genomes every million years. Most genes belong to larger gene families of shared ancestry, known as homology. Novel genes are produced by several methods, commonly through the duplication and mutation of an ancestral gene, or by recombining parts of different genes to form new combinations with new functions.

Here, protein domains act as modules, each with a particular and independent function, that can be mixed together to produce genes encoding new proteins with novel properties. For example, the human eye uses four genes to make structures that sense light: three for cone cell or color vision and one for rod cell or night vision; all four arose from a single ancestral gene. Another advantage of duplicating a gene (or even an entire genome) is that this increases engineering redundancy; this allows one gene in the pair to acquire a new function while the other copy performs the original function. Other types of mutation occasionally create new genes from previously noncoding DNA.

Changes in chromosome number may involve even larger mutations, where segments of the DNA within chromosomes break and then rearrange. For example, in the Homininae, two chromosomes fused to produce human chromosome 2; this fusion did not occur in the lineage of the other apes, and they retain these separate chromosomes. In evolution, the most important role of such chromosomal rearrangements may be to accelerate the divergence of a population into new species by making populations less likely to interbreed, thereby preserving genetic differences between these populations.

Sequences of DNA that can move about the genome, such as transposons, make up a major fraction of the genetic material of plants and animals, and may have been important in the evolution of genomes. For example, more than a million copies of the Alu sequence are present in the human genome, and these sequences have now been recruited to perform functions such as regulating gene expression. Another effect of these mobile DNA sequences is that when they move within a genome, they can mutate or delete existing genes and thereby produce genetic diversity.

Nonlethal mutations accumulate within the gene pool and increase the amount of genetic variation. The abundance of some genetic changes within the gene pool can be reduced by natural selection, while other "more favorable" mutations may accumulate and result in adaptive changes.

Prodryas persephone, a Late Eocene butterfly

For example, a butterfly may produce offspring with new mutations. The majority of these mutations will have no effect; but one might change the color of one of the butterfly's offspring, making it harder (or easier) for predators to see. If this color change is advantageous, the chance of this butterfly's surviving and producing its own offspring are a little better, and over time the number of butterflies with this mutation may form a larger percentage of the population.

Neutral mutations are defined as mutations whose effects do not influence the fitness of an individual. These can accumulate over time due to genetic drift. It is believed that the overwhelming majority of mutations have no significant effect on an organism's fitness. Also, DNA repair mechanisms are able to mend most changes before they become permanent mutations, and many organisms have mechanisms for eliminating otherwise-permanently mutated somatic cells.

Beneficial mutations can improve reproductive success.

Causes

Four classes of mutations are (1) spontaneous mutations (molecular decay), (2) mutations due to error-prone replication bypass of naturally occurring DNA damage (also called error-prone translesion synthesis), (3) errors introduced during DNA repair, and (4) induced mutations caused by mutagens. Scientists may also deliberately introduce mutant sequences through DNA manipulation for the sake of scientific experimentation.

Spontaneous Mutation

Spontaneous mutations on the molecular level can be caused by:

- Tautomerism — A base is changed by the repositioning of a hydrogen atom, altering the hydrogen bonding pattern of that base, resulting in incorrect base pairing during replication.
- Depurination — Loss of a purine base (A or G) to form an apurinic site (AP site).
- Deamination — Hydrolysis changes a normal base to an atypical base containing a keto group in place of the original amine group. Examples include C → U and A → HX (hypoxanthine), which can be corrected by DNA repair mechanisms; and 5MeC (5-methylcytosine) → T, which is less likely to be detected as a mutation because thymine is a normal DNA base.
- Slipped strand mispairing — Denaturation of the new strand from the template during replication, followed by renaturation in a different spot ("slipping"). This can lead to insertions or deletions.

Error-prone Replication Bypass

There is increasing evidence that the majority of spontaneously arising mutations are due to error-prone replication (translesion synthesis) past DNA damage in the template strand. Naturally occurring oxidative DNA damages arise at least 10,000 times per cell per day in humans and 50,000 times or more per cell per day in rats. In mice, the majority of mutations are caused by translesion synthesis. Likewise, in yeast, Kunz et al. found that more than 60% of the spontaneous single base pair substitutions and deletions were caused by translesion synthesis.

Errors Introduced During DNA Repair

Although naturally occurring double-strand breaks occur at a relatively low frequency in DNA, their repair often causes mutation. Non-homologous end joining (NHEJ) is a major pathway for repairing double-strand breaks. NHEJ involves removal of a few nucleotides to allow somewhat inaccurate alignment of the two ends for rejoining followed by addition of nucleotides to fill in gaps. As a consequence, NHEJ often introduces mutations.

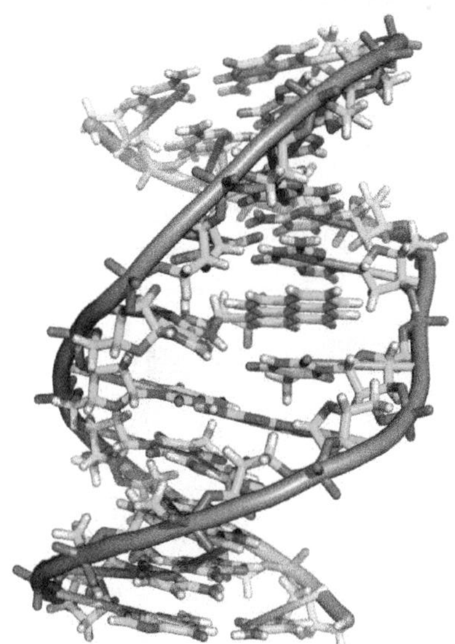

A covalent adduct between the metabolite of benzo[*a*]pyrene, the major mutagen in tobacco smoke, and DNA

Induced Mutation

Induced mutations on the molecular level can be caused by:-

- Chemicals
  - Hydroxylamine
  - Base analogs (e.g., Bromodeoxyuridine (BrdU))
  - Alkylating agents (e.g., *N*-ethyl-*N*-nitrosourea (ENU)). These agents can mutate both replicating and non-replicating DNA. In contrast, a base analog can mutate the DNA only when the analog is incorporated in replicating the DNA. Each of these classes of chemical mutagens has certain effects that then lead to transitions, transversions, or deletions.
  - Agents that form DNA adducts (e.g., ochratoxin A)
  - DNA intercalating agents (e.g., ethidium bromide)
  - DNA crosslinkers
  - Oxidative damage
  - Nitrous acid converts amine groups on A and C to diazo groups, altering their hydrogen bonding patterns, which leads to incorrect base pairing during replication.
- Radiation
  - Ultraviolet light (UV) (non-ionizing radiation). Two nucleotide bases in DNA—cytosine and thymine—are most vulnerable to radiation that can change their properties. UV light can induce adjacent pyrimidine bases in a DNA strand to become covalently joined as a pyrimidine dimer. UV radiation, in particular longer-wave UVA, can also cause oxidative damage to DNA.

Classification of Mutation Types

By Effect on Structure

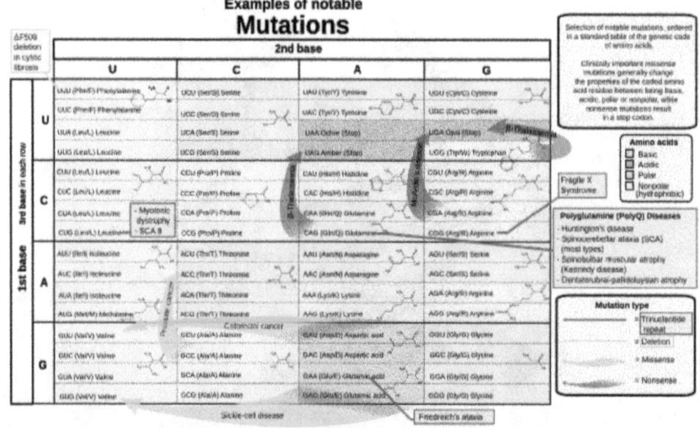

Selection of disease-causing mutations, in a standard table of the genetic code of amino acids.

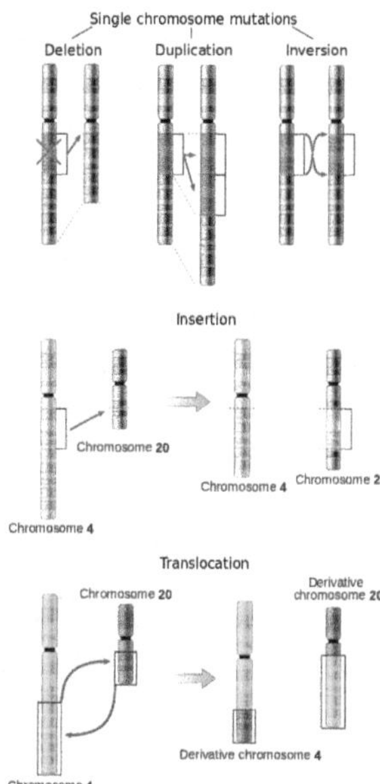

Illustrations of five types of chromosomal mutations.

The sequence of a gene can be altered in a number of ways. Gene mutations have varying effects on health depending on where they occur and whether they alter the function of essential proteins. Mutations in the structure of genes can be classified as:

- Small-scale mutations, such as those affecting a small gene in one or a few nucleotides, including:
    - Substitution mutations, often caused by chemicals or malfunction of DNA replication, exchange a single nucleotide for another. These changes are classified as transitions or transversions. Most common is the transition that exchanges a purine for a purine (A ↔ G) or a pyrimidine for a pyrimidine, (C ↔ T). A transition can be caused by nitrous acid, base mis-pairing, or mutagenic base analogs such as BrdU. Less common is a transversion, which exchanges a purine for a pyrimidine or a pyrimidine for a purine (C/T ↔ A/G). An example of a transversion is the conversion of adenine (A) into a cytosine (C). A point mutation can be reversed by another point mutation, in which the nucleotide is changed back to its original state (true reversion) or by second-site reversion (a complementary mutation elsewhere that results in regained gene functionality). Point mutations that occur within the protein coding region of a gene may be classified into three kinds, depending upon what the erroneous codon codes for:
        - Silent mutations, which code for the same (or a sufficiently similar) amino acid.
        - Missense mutations, which code for a different amino acid.
        - Nonsense mutations, which code for a stop codon and can truncate the protein.
    - Insertions add one or more extra nucleotides into the DNA. They are usually caused by transposable elements, or errors during replication of repeating elements.

Insertions in the coding region of a gene may alter splicing of the mRNA (splice site mutation), or cause a shift in the reading frame (frameshift), both of which can significantly alter the gene product. Insertions can be reversed by excision of the transposable element.

- Deletions remove one or more nucleotides from the DNA. Like insertions, these mutations can alter the reading frame of the gene. In general, they are irreversible: Though exactly the same sequence might in theory be restored by an insertion, transposable elements able to revert a very short deletion (say 1–2 bases) in *any* location either are highly unlikely to exist or do not exist at all.

- Large-scale mutations in chromosomal structure, including:
  - Amplifications (or gene duplications) leading to multiple copies of all chromosomal regions, increasing the dosage of the genes located within them.
  - Deletions of large chromosomal regions, leading to loss of the genes within those regions.
  - Mutations whose effect is to juxtapose previously separate pieces of DNA, potentially bringing together separate genes to form functionally distinct fusion genes (e.g., bcr-abl). These include:
    - Chromosomal translocations: interchange of genetic parts from nonhomologous chromosomes.
    - Interstitial deletions: an intra-chromosomal deletion that removes a segment of DNA from a single chromosome, thereby apposing previously distant genes. For example, cells isolated from a human astrocytoma, a type of brain tumor, were found to have a chromosomal deletion removing sequences between the Fused in Glioblastoma (FIG) gene and the receptor tyrosine kinase (ROS), producing a fusion protein (FIG-ROS). The abnormal FIG-ROS fusion protein has constitutively active kinase activity that causes oncogenic transformation (a transformation from normal cells to cancer cells).
    - Chromosomal inversions: reversing the orientation of a chromosomal segment.
  - Loss of heterozygosity: loss of one allele, either by a deletion or a genetic recombination event, in an organism that previously had two different alleles.

By Effect on Function

- Loss-of-function mutations, also called inactivating mutations, result in the gene product having less or no function (being partially or wholly inactivated). When the allele has a complete loss of function (null allele), it is often called an amorphic mutation in the Muller's morphs schema. Phenotypes associated with such mutations are most often recessive. Exceptions are when the organism is haploid, or when the reduced dosage of a normal gene product is not enough for a normal phenotype (this is called haploinsufficiency).

- Gain-of-function mutations, also called activating mutations, change the gene product such that its effect gets stronger (enhanced activation) or even is superseded by a different and abnormal function. When the new allele is created, a heterozygote containing the newly created allele as well as the original will express the new allele; genetically this defines the mutations as dominant phenotypes. Often called a neomorphic mutation.
- Dominant negative mutations (also called antimorphic mutations) have an altered gene product that acts antagonistically to the wild-type allele. These mutations usually result in an altered molecular function (often inactive) and are characterized by a dominant or semi-dominant phenotype. In humans, dominant negative mutations have been implicated in cancer (e.g., mutations in genes p53, ATM, CEBPA and PPARgamma). Marfan syndrome is caused by mutations in the *FBN1* gene, located on chromosome 15, which encodes fibrillin-1, a glycoprotein component of the extracellular matrix. Marfan syndrome is also an example of dominant negative mutation and haploinsufficiency.
- Lethal mutations are mutations that lead to the death of the organisms that carry the mutations.
- A back mutation or reversion is a point mutation that restores the original sequence and hence the original phenotype.

By Effect on Fitness

In applied genetics, it is usual to speak of mutations as either harmful or beneficial.

- A harmful, or deleterious, mutation decreases the fitness of the organism.
- A beneficial, or advantageous mutation increases the fitness of the organism. Mutations that promotes traits that are desirable, are also called beneficial. In theoretical population genetics, it is more usual to speak of mutations as deleterious or advantageous than harmful or beneficial.
- A neutral mutation has no harmful or beneficial effect on the organism. Such mutations occur at a steady rate, forming the basis for the molecular clock. In the neutral theory of molecular evolution, neutral mutations provide genetic drift as the basis for most variation at the molecular level.
- A nearly neutral mutation is a mutation that may be slightly deleterious or advantageous, although most nearly neutral mutations are slightly deleterious.

Distribution of Fitness Effects

Attempts have been made to infer the distribution of fitness effects (DFE) using mutagenesis experiments and theoretical models applied to molecular sequence data. DFE, as used to determine the relative abundance of different types of mutations (i.e., strongly deleterious, nearly neutral or advantageous), is relevant to many evolutionary questions, such as the maintenance of genetic variation, the rate of genomic decay, the maintenance of outcrossing sexual reproduction as opposed to inbreeding and the evolution of sex and genetic recombination. In summary, the DFE plays an important role in predicting evolutionary dynamics. A variety of approaches have been used to study the DFE, including theoretical, experimental and analytical methods.

- Mutagenesis experiment: The direct method to investigate the DFE is to induce mutations and then measure the mutational fitness effects, which has already been done in viruses, bacteria, yeast, and *Drosophila*. For example, most studies of the DFE in viruses used site-directed mutagenesis to create point mutations and measure relative fitness of each mutant. In *Escherichia coli*, one study used transposon mutagenesis to directly measure the fitness of a random insertion of a derivative of Tn10. In yeast, a combined mutagenesis and deep sequencing approach has been developed to generate high-quality systematic mutant libraries and measure fitness in high throughput. However, given that many mutations have effects too small to be detected and that mutagenesis experiments can detect only mutations of moderately large effect; DNA sequence data analysis can provide valuable information about these mutations.

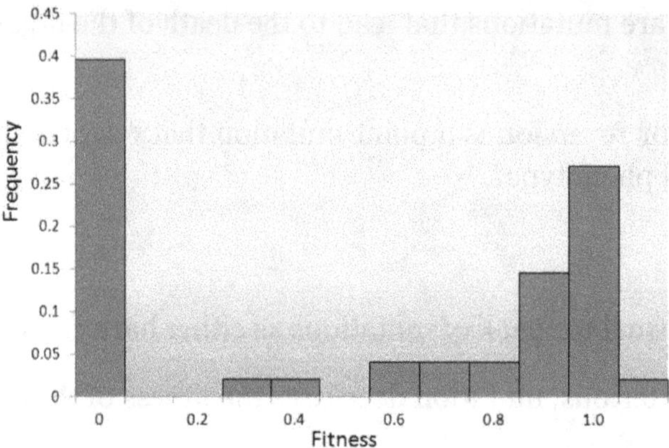

The distribution of fitness effects (DFE) of mutations in vesicular stomatitis virus. In this experiment, random mutations were introduced into the virus by site-directed mutagenesis, and the fitness of each mutant was compared with the ancestral type. A fitness of zero, less than one, one, more than one, respectively, indicates that mutations are lethal, deleterious, neutral, and advantageous.

- Molecular sequence analysis: With rapid development of DNA sequencing technology, an enormous amount of DNA sequence data is available and even more is forthcoming in the future. Various methods have been developed to infer the DFE from DNA sequence data. By examining DNA sequence differences within and between species, we are able to infer various characteristics of the DFE for neutral, deleterious and advantageous mutations. To be specific, the DNA sequence analysis approach allows us to estimate the effects of mutations with very small effects, which are hardly detectable through mutagenesis experiments.

One of the earliest theoretical studies of the distribution of fitness effects was done by Motoo Kimura, an influential theoretical population geneticist. His neutral theory of molecular evolution proposes that most novel mutations will be highly deleterious, with a small fraction being neutral. Hiroshi Akashi more recently proposed a bimodal model for the DFE, with modes centered around highly deleterious and neutral mutations. Both theories agree that the vast majority of novel mutations are neutral or deleterious and that advantageous mutations are rare, which has been supported by experimental results. One example is a study done on the DFE of random mutations in vesicular stomatitis virus. Out of all mutations, 39.6% were lethal, 31.2% were non-lethal deleterious, and 27.1% were neutral. Another example comes from a high throughput mutagenesis experiment with yeast. In this experiment it was shown that the overall DFE is bimodal, with a cluster of neutral mutations, and a broad distribution of deleterious mutations.

Though relatively few mutations are advantageous, those that are play an important role in evolutionary changes. Like neutral mutations, weakly selected advantageous mutations can be lost due to random genetic drift, but strongly selected advantageous mutations are more likely to be fixed. Knowing the DFE of advantageous mutations may lead to increased ability to predict the evolutionary dynamics. Theoretical work on the DFE for advantageous mutations has been done by John H. Gillespie and H. Allen Orr. They proposed that the distribution for advantageous mutations should be exponential under a wide range of conditions, which, in general, has been supported by experimental studies, at least for strongly selected advantageous mutations.

In general, it is accepted that the majority of mutations are neutral or deleterious, with rare mutations being advantageous; however, the proportion of types of mutations varies between species. This indicates two important points: first, the proportion of effectively neutral mutations is likely to vary between species, resulting from dependence on effective population size; second, the average effect of deleterious mutations varies dramatically between species. In addition, the DFE also differs between coding regions and noncoding regions, with the DFE of noncoding DNA containing more weakly selected mutations.

By Impact on Protein Sequence

- A frameshift mutation is a mutation caused by insertion or deletion of a number of nucleotides that is not evenly divisible by three from a DNA sequence. Due to the triplet nature of gene expression by codons, the insertion or deletion can disrupt the reading frame, or the grouping of the codons, resulting in a completely different translation from the original. The earlier in the sequence the deletion or insertion occurs, the more altered the protein produced is.

  In contrast, any insertion or deletion that is evenly divisible by three is termed an *in-frame mutation*

- A nonsense mutation is a point mutation in a sequence of DNA that results in a premature stop codon, or a *nonsense codon* in the transcribed mRNA, and possibly a truncated, and often nonfunctional protein product.

- Missense mutations or *nonsynonymous mutations* are types of point mutations where a single nucleotide is changed to cause substitution of a different amino acid. This in turn can render the resulting protein nonfunctional. Such mutations are responsible for diseases such as Epidermolysis bullosa, sickle-cell disease, and SOD1-mediated ALS.

- A neutral mutation is a mutation that occurs in an amino acid codon that results in the use of a different, but chemically similar, amino acid. The similarity between the two is enough that little or no change is often rendered in the protein. For example, a change from AAA to AGA will encode arginine, a chemically similar molecule to the intended lysine.

- Silent mutations are mutations that do not result in a change to the amino acid sequence of a protein, unless the changed amino acid is sufficiently similar to the original. They may occur in a region that does not code for a protein, or they may occur within a codon in a manner that does not alter the final amino acid sequence. The phrase *silent mutation* is often used interchangeably with the phrase *synonymous mutation*; however, synonymous mutations are a subcategory of the former, occurring only within exons (and necessarily exactly preserving the amino acid sequence of the protein). Synonymous mutations occur due to the degenerate nature of the genetic code.

By Inheritance

In multicellular organisms with dedicated reproductive cells, mutations can be subdivided into germline mutations, which can be passed on to descendants through their reproductive cells, and somatic mutations (also called acquired mutations), which involve cells outside the dedicated reproductive group and which are not usually transmitted to descendants.

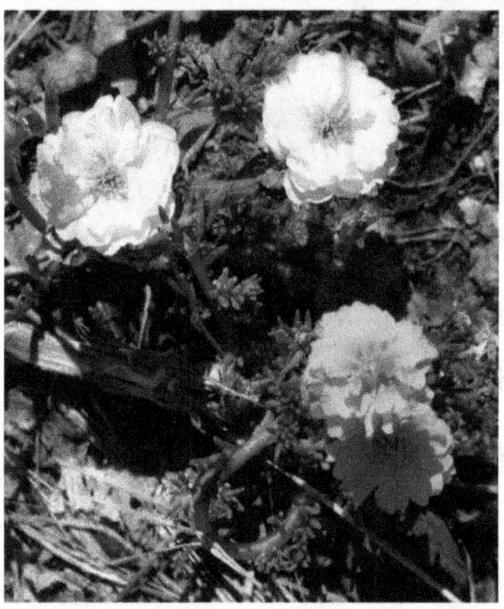

A mutation has caused this garden moss rose to produce flowers of different colors. This is a somatic mutation that may also be passed on in the germline.

A germline mutation gives rise to a *constitutional mutation* in the offspring, that is, a mutation that is present in every cell. A constitutional mutation can also occur very soon after fertilisation, or continue from a previous constitutional mutation in a parent.

The distinction between germline and somatic mutations is important in animals that have a dedicated germline to produce reproductive cells. However, it is of little value in understanding the effects of mutations in plants, which lack dedicated germline. The distinction is also blurred in those animals that reproduce asexually through mechanisms such as budding, because the cells that give rise to the daughter organisms also give rise to that organism's germline. A new germline mutation that was not inherited from either parent is called a *de novo* mutation.

Diploid organisms (e.g., humans) contain two copies of each gene—a paternal and a maternal allele. Based on the occurrence of mutation on each chromosome, we may classify mutations into three types.

- A heterozygous mutation is a mutation of only one allele.
- A homozygous mutation is an identical mutation of both the paternal and maternal alleles.
- Compound heterozygous mutations or a genetic compound consists of two different mutations in the paternal and maternal alleles.

A wild type or homozygous non-mutated organism is one in which neither allele is mutated.

Special Classes

- Conditional mutation is a mutation that has wild-type (or less severe) phenotype under certain "permissive" environmental conditions and a mutant phenotype under certain "restrictive" conditions. For example, a temperature-sensitive mutation can cause cell death at high temperature (restrictive condition), but might have no deleterious consequences at a lower temperature (permissive condition).
- Replication timing quantitative trait loci affects DNA replication.

Nomenclature

In order to categorize a mutation as such, the "normal" sequence must be obtained from the DNA of a "normal" or "healthy" organism (as opposed to a "mutant" or "sick" one), it should be identified and reported; ideally, it should be made publicly available for a straightforward nucleotide-by-nucleotide comparison, and agreed upon by the scientific community or by a group of expert geneticists and biologists, who have the responsibility of establishing the *standard* or so-called "consensus" sequence. This step requires a tremendous scientific effort. Once the consensus sequence is known, the mutations in a genome can be pinpointed, described, and classified. The committee of the Human Genome Variation Society (HGVS) has developed the standard human sequence variant nomenclature, which should be used by researchers and DNA diagnostic centers to generate unambiguous mutation descriptions. In principle, this nomenclature can also be used to describe mutations in other organisms. The nomenclature specifies the type of mutation and base or amino acid changes.

- Nucleotide substitution (e.g., 76A>T) — The number is the position of the nucleotide from the 5' end; the first letter represents the wild-type nucleotide, and the second letter represents the nucleotide that replaced the wild type. In the given example, the adenine at the 76th position was replaced by a thymine.
    - If it becomes necessary to differentiate between mutations in genomic DNA, mitochondrial DNA, and RNA, a simple convention is used. For example, if the 100th base of a nucleotide sequence mutated from G to C, then it would be written as g.100G>C if the mutation occurred in genomic DNA, m.100G>C if the mutation occurred in mitochondrial DNA, or r.100g>c if the mutation occurred in RNA. Note that, for mutations in RNA, the nucleotide code is written in lower case.
- Amino acid substitution (e.g., D111E) — The first letter is the one letter code of the wild-type amino acid, the number is the position of the amino acid from the N-terminus, and the second letter is the one letter code of the amino acid present in the mutation. Nonsense mutations are represented with an X for the second amino acid (e.g. D111X).

- Amino acid deletion (e.g., ΔF508) — The Greek letter Δ (delta) indicates a deletion. The letter refers to the amino acid present in the wild type and the number is the position from the N terminus of the amino acid were it to be present as in the wild type.

Mutation Rates

Mutation rates vary substantially across species, and the evolutionary forces that generally determine mutation are the subject of ongoing investigation.

Harmful Mutations

Changes in DNA caused by mutation can cause errors in protein sequence, creating partially or completely non-functional proteins. Each cell, in order to function correctly, depends on thousands of proteins to function in the right places at the right times. When a mutation alters a protein that plays a critical role in the body, a medical condition can result. A condition caused by mutations in one or more genes is called a Sheik disorder. Some mutations alter a gene's DNA base sequence but do not change the function of the protein made by the gene. One study on the comparison of genes between different species of *Drosophila* suggests that if a mutation does change a protein, this will probably be harmful, with an estimated 70 percent of amino acid polymorphisms having damaging effects, and the remainder being either neutral or weakly beneficial.]] Studies have shown that only 7% of point mutations in noncoding DNA of yeast are deleterious and 12% in coding DNA are deleterious. The rest of the mutations are either neutral or slightly beneficial.

If a mutation is present in a germ cell, it can give rise to offspring that carries the mutation in all of its cells. This is the case in hereditary diseases. In particular, if there is a mutation in a DNA repair gene within a germ cell, humans carrying such germline mutations may have an increased risk of cancer. An example of one is albinism, a mutation that occurs in the OCA1 or OCA2 gene. Individuals with this disorder are more prone to many types of cancers, other disorders and have impaired vision. On the other hand, a mutation may occur in a somatic cell of an organism. Such mutations will be present in all descendants of this cell within the same organism, and certain mutations can cause the cell to become malignant, and, thus, cause cancer.

A DNA damage can cause an error when the DNA is replicated, and this error of replication can cause a gene mutation that, in turn, could cause a genetic disorder. DNA damages are repaired by the DNA repair system of the cell. Each cell has a number of pathways through which enzymes recognize and repair damages in DNA. Because DNA can be damaged in many ways, the process of DNA repair is an important way in which the body protects itself from disease. Once DNA damage has given rise to a mutation, the mutation cannot be repaired. DNA repair pathways can only recognize and act on "abnormal" structures in the DNA. Once a mutation occurs in a gene sequence it then has normal DNA structure and cannot be repaired.

Beneficial Mutations

Although mutations that cause changes in protein sequences can be harmful to an organism, on occasions the effect may be positive in a given environment. In this case, the mutation may enable the mutant organism to withstand particular environmental stresses better than wild-type organisms, or reproduce more quickly. In these cases a mutation will tend to become more common in a population through natural selection.

For example, a specific 32 base pair deletion in human CCR5 (CCR5-Δ32) confers HIV resistance to homozygotes and delays AIDS onset in heterozygotes. One possible explanation of the etiology of the relatively high frequency of CCR5-Δ32 in the European population is that it conferred resistance to the bubonic plague in mid-14th century Europe. People with this mutation were more likely to survive infection; thus its frequency in the population increased. This theory could explain why this mutation is not found in Southern Africa, which remained untouched by bubonic plague. A newer theory suggests that the selective pressure on the CCR5 Delta 32 mutation was caused by smallpox instead of the bubonic plague.

An example of a harmful mutation is sickle-cell disease, a blood disorder in which the body produces an abnormal type of the oxygen-carrying substance hemoglobin in the red blood cells. One-third of all indigenous inhabitants of Sub-Saharan Africa carry the gene, because, in areas where malaria is common, there is a survival value in carrying only a single sickle-cell gene (sickle cell trait). Those with only one of the two alleles of the sickle-cell disease are more resistant to malaria, since the infestation of the malaria *Plasmodium* is halted by the sickling of the cells that it infests.

Prion Mutations

Prions are proteins and do not contain genetic material. However, prion replication has been shown to be subject to mutation and natural selection just like other forms of replication. The human gene PRNP codes for the major prion protein, PrP, and is subject to mutations that can give rise to disease-causing prions.

Somatic Mutations

A change in the genetic structure that is not inherited from a parent, and also not passed to offspring, is called a *somatic cell genetic mutation* or *acquired mutation*.

Cells with heterozygous mutations (one good copy of gene and one mutated copy) may function normally with the unmutated copy until the good copy has been spontaneously somatically mutated. This kind of mutation happens all the time in living organisms, but it is difficult to measure the rate. Measuring this rate is important in predicting the rate at which people may develop cancer.

Point mutations may arise from spontaneous mutations that occur during DNA replication. The rate of mutation may be increased by mutagens. Mutagens can be physical, such as radiation from UV rays, X-rays or extreme heat, or chemical (molecules that misplace base pairs or disrupt the helical shape of DNA). Mutagens associated with cancers are often studied to learn about cancer and its prevention.

DNA

Deoxyribonucleic acid is a molecule that carries the genetic instructions used in the growth, development, functioning and reproduction of all known living organisms and many viruses. DNA and

RNA are nucleic acids; alongside proteins, lipids and complex carbohydrates (polysaccharides), they are one of the four major types of macromolecules that are essential for all known forms of life. Most DNA molecules consist of two biopolymer strands coiled around each other to form a double helix.

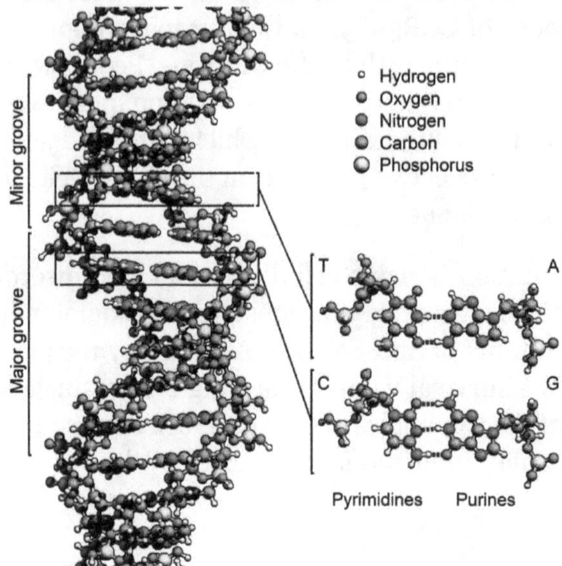

The structure of the DNA double helix. The atoms in the structure are colour-coded by element and the detailed structure of two base pairs are shown in the bottom right.

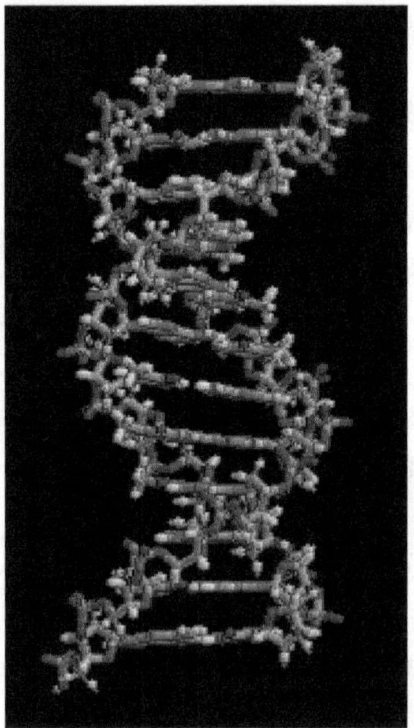

The structure of part of a DNA double helix

The two DNA strands are termed polynucleotides since they are composed of simpler monomer units called nucleotides. Each nucleotide is composed of one of four nitrogen-containing nucleobases—either cytosine (C), guanine (G), adenine (A), or thymine (T)—and a sugar called deoxyribose and a phosphate group. The nucleotides are joined to one another in a chain by covalent bonds between the sugar of one nucleotide and the phosphate of the next, resulting in an alternating sugar-phosphate backbone. The nitrogenous bases of the two separate polynucleotide strands are bound together (according to base pairing rules (A with T, and C with G) with hydrogen bonds to make double-stranded DNA. The total amount of related DNA base pairs on Earth is estimated at 5.0×10^{37}, and weighs 50 billion tonnes. In comparison, the total mass of the biosphere has been estimated to be as much as 4 trillion tons of carbon (TtC).

DNA stores biological information. The DNA backbone is resistant to cleavage, and both strands of the double-stranded structure store the same biological information. This information is replicated as and when the two strands separate. A large part of DNA (more than 98% for humans) is non-coding, meaning that these sections do not serve as patterns for protein sequences.

The two strands of DNA run in opposite directions to each other and are thus antiparallel. Attached to each sugar is one of four types of nucleobases (informally, *bases*). It is the sequence of these four nucleobases along the backbone that encodes biological information. RNA strands are created using DNA strands as a template in a process called transcription. Under the genetic code, these RNA strands are translated to specify the sequence of amino acids within proteins in a process called translation.

Within eukaryotic cells, DNA is organized into long structures called chromosomes. During cell division these chromosomes are duplicated in the process of DNA replication, providing each cell its own complete set of chromosomes. Eukaryotic organisms (animals, plants, fungi, and protists) store most of their DNA inside the cell nucleus and some of their DNA in organelles, such as mitochondria or chloroplasts. In contrast, prokaryotes (bacteria and archaea) store their DNA only in the cytoplasm. Within the eukaryotic chromosomes, chromatin proteins such as histones compact and organize DNA. These compact structures guide the interactions between DNA and other proteins, helping control which parts of the DNA are transcribed.

DNA was first isolated by Friedrich Miescher in 1869. Its molecular structure was identified by James Watson and Francis Crick in 1953, whose model-building efforts were guided by X-ray diffraction data acquired by Rosalind Franklin. DNA is used by researchers as a molecular tool to explore physical laws and theories, such as the ergodic theorem and the theory of elasticity. The unique material properties of DNA have made it an attractive molecule for material scientists and engineers interested in micro- and nano-fabrication. Among notable advances in this field are DNA origami and DNA-based hybrid materials.

Properties

DNA is a long polymer made from repeating units called nucleotides. The structure of DNA is non-static, all species comprises two helical chains each coiled round the same axis, and each with a pitch of 34 ångströms (3.4 nanometres) and a radius of 10 ångströms (1.0 nanometre). According to another study, when measured in a particular solution, the DNA chain measured 22 to 26 ångströms wide (2.2 to 2.6 nanometres), and one nucleotide unit measured 3.3 Å (0.33 nm) long. Although each individual repeating unit is very small, DNA polymers can be very large molecules containing millions of nucleotides. For instance, the DNA in the largest human chromosome, chromosome number 1, consists of approximately 220 million base pairs and would be 85 mm long if straightened.

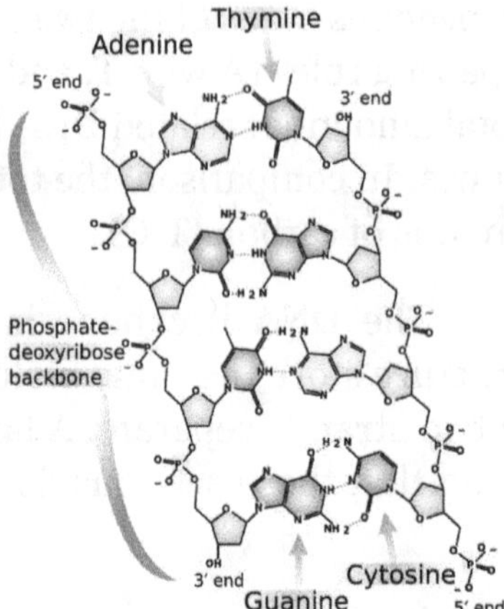

Chemical structure of DNA; hydrogen bonds shown as dotted lines

In living organisms DNA does not usually exist as a single molecule, but instead as a pair of molecules that are held tightly together. These two long strands entwine like vines, in the shape of a double helix. The nucleotide contains both a segment of the backbone of the molecule (which holds the chain together) and a nucleobase (which interacts with the other DNA strand in the helix). A nucleobase linked to a sugar is called a nucleoside and a base linked to a sugar and one or more phosphate groups is called a nucleotide. A polymer comprising multiple linked nucleotides (as in DNA) is called a polynucleotide.

The backbone of the DNA strand is made from alternating phosphate and sugar residues. The sugar in DNA is 2-deoxyribose, which is a pentose (five-carbon) sugar. The sugars are joined together by phosphate groups that form phosphodiester bonds between the third and fifth carbon atoms of adjacent sugar rings. These asymmetric bonds mean a strand of DNA has a direction. In a double helix, the direction of the nucleotides in one strand is opposite to their direction in the other strand: the strands are *antiparallel*. The asymmetric ends of DNA strands are said to have a directionality of *five prime* (5′) and *three prime* (3′), with the 5′ end having a terminal phosphate group and the 3′ end a terminal hydroxyl group. One major difference between DNA and RNA is the sugar, with the 2-deoxyribose in DNA being replaced by the alternative pentose sugar ribose in RNA.

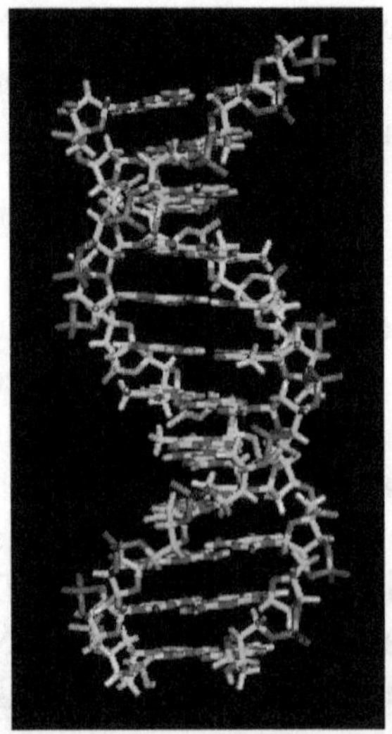

A section of DNA. The bases lie horizontally between the two spiraling strands. (animated version).

The DNA double helix is stabilized primarily by two forces: hydrogen bonds between nucleotides and base-stacking interactions among aromatic nucleobases. In the aqueous environment of the cell, the conjugated π bonds of nucleotide bases align perpendicular to the axis of the DNA molecule, minimizing their interaction with the solvation shell. The four bases found in DNA are adenine (A), cytosine (C), guanine (G) and thymine (T). These four bases are attached to the sugar-phosphate to form the complete nucleotide, as shown for adenosine monophosphate. Adenine pairs with thymine and guanine pairs with cytosine. It was represented by A-T base pairs and G-C base pairs.

Nucleobase Classification

The nucleobases are classified into two types: the purines, A and G, being fused five- and six-membered heterocyclic compounds, and the pyrimidines, the six-membered rings C and T. A fifth pyrimidine nucleobase, uracil (U), usually takes the place of thymine in RNA and differs from thymine by lacking a methyl group on its ring. In addition to RNA and DNA, many artificial nucleic acid analogues have been created to study the properties of nucleic acids, or for use in biotechnology.

Uracil is not usually found in DNA, occurring only as a breakdown product of cytosine. However, in several bacteriophages, *Bacillus subtilis* bacteriophages PBS1 and PBS2 and *Yersinia* bacteriophage piR1-37, thymine has been replaced by uracil. Another phage - Staphylococcal phage S6 - has been identified with a genome where thymine has been replaced by uracil.

Base J (beta-d-glucopyranosyloxymethyluracil), a modified form of uracil, is also found in several organisms: the flagellates *Diplonema* and *Euglena*, and all the kinetoplastid genera. Biosynthesis of J occurs in two steps: in the first step a specific thymidine in DNA is converted into hydroxymethyldeoxyuridine; in the second HOMedU is glycosylated to form J. Proteins that bind specifically to this base have been identified. These proteins appear to be distant relatives of the Tet1 oncogene that is involved in the pathogenesis of acute myeloid leukemia. J appears to act as a termination signal for RNA polymerase II.

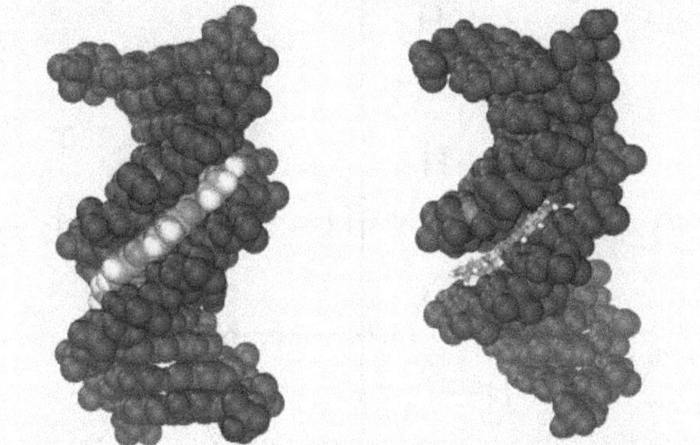

DNA major and minor grooves. The later is a binding site for the Hoechst stain dye 33258.

Grooves

Twin helical strands form the DNA backbone. Another double helix may be found tracing the spaces, or grooves, between the strands. These voids are adjacent to the base pairs and may provide a binding site. As the strands are not symmetrically located with respect to each other, the grooves are unequally sized. One groove, the major groove, is 22 Å wide and the other, the minor groove, is 12 Å wide. The width of the major groove means that the edges of the bases are more accessible in the major groove than in the minor groove. As a result, proteins such as transcription factors that can bind to specific sequences in double-stranded DNA usually make contact with the sides of the bases exposed in the major groove. This situation varies in unusual conformations of DNA within the cell, but the major and minor grooves are always named to reflect the differences in size that would be seen if the DNA is twisted back into the ordinary B form.

Base Pairing

In a DNA double helix, each type of nucleobase on one strand bonds with just one type of nucleobase on the other strand. This is called complementary base pairing. Here, purines form hydrogen bonds to pyrimidines, with adenine bonding only to thymine in two hydrogen bonds, and cytosine bonding only to guanine in three hydrogen bonds. This arrangement of two nucleotides binding together across the double helix is called a base pair. As hydrogen bonds are not covalent, they can be broken and rejoined relatively easily. The two strands of DNA in a double helix can thus be pulled apart like a zipper, either by a mechanical force or high temperature. As a result of this base pair complementarity, all the information in the double-stranded sequence of a DNA helix is duplicated on each strand, which is vital in DNA replication. This reversible and specific interaction between complementary base pairs is critical for all the functions of DNA in living organisms.

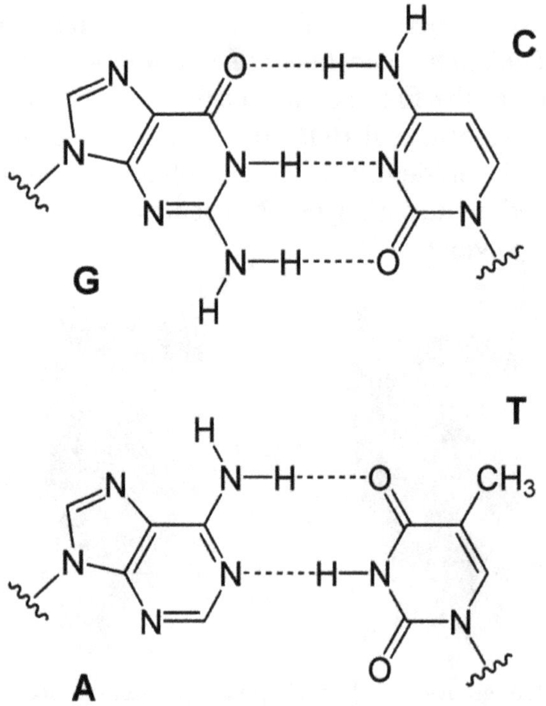

Top, a GC base pair with three hydrogen bonds. Bottom, an AT base pair with two hydrogen bonds. Non-covalent hydrogen bonds between the pairs are shown as dashed lines.

The two types of base pairs form different numbers of hydrogen bonds, AT forming two hydrogen bonds, and GC forming three hydrogen bonds. DNA with high GC-content is more stable than DNA with low GC-content.

As noted above, most DNA molecules are actually two polymer strands, bound together in a helical fashion by noncovalent bonds; this double stranded structure (dsDNA) is maintained largely by the intrastrand base stacking interactions, which are strongest for G,C stacks. The two strands can come apart – a process known as melting – to form two single-stranded DNA molecules (ssDNA) molecules. Melting occurs at high temperature, low salt and high pH (low pH also melts DNA, but since DNA is unstable due to acid depurination, low pH is rarely used).

The stability of the dsDNA form depends not only on the GC-content (% G,C basepairs) but also on sequence (since stacking is sequence specific) and also length (longer molecules are more stable). The stability can be measured in various ways; a common way is the "melting temperature", which is the temperature at which 50% of the ds molecules are converted to ss molecules; melting temperature is dependent on ionic strength and the concentration of DNA. As a result, it is both the percentage of GC base pairs and the overall length of a DNA double helix that determines the strength of the association between the two strands of DNA. Long DNA helices with a high GC-content have stronger-interacting strands, while short helices with high AT content have weaker-interacting strands. In biology, parts of the DNA double helix that need to separate easily, such as the TATAAT Pribnow box in some promoters, tend to have a high AT content, making the strands easier to pull apart.

In the laboratory, the strength of this interaction can be measured by finding the temperature necessary to break the hydrogen bonds, their melting temperature (also called T_m value). When all the base pairs in a DNA double helix melt, the strands separate and exist in solution as two entirely independent molecules. These single-stranded DNA molecules (*ssDNA*) have no single common shape, but some conformations are more stable than others.

Sense and Antisense

A DNA sequence is called "sense" if its sequence is the same as that of a messenger RNA copy that is translated into protein. The sequence on the opposite strand is called the "antisense" sequence. Both sense and antisense sequences can exist on different parts of the same strand of DNA (i.e. both strands can contain both sense and antisense sequences). In both prokaryotes and eukaryotes, antisense RNA sequences are produced, but the functions of these RNAs are not entirely clear. One proposal is that antisense RNAs are involved in regulating gene expression through RNA-RNA base pairing.

A few DNA sequences in prokaryotes and eukaryotes, and more in plasmids and viruses, blur the distinction between sense and antisense strands by having overlapping genes. In these cases, some DNA sequences do double duty, encoding one protein when read along one strand, and a second protein when read in the opposite direction along the other strand. In bacteria, this overlap may be involved in the regulation of gene transcription, while in viruses, overlapping genes increase the amount of information that can be encoded within the small viral genome.

Supercoiling

DNA can be twisted like a rope in a process called DNA supercoiling. With DNA in its "relaxed" state, a strand usually circles the axis of the double helix once every 10.4 base pairs, but if the DNA is twisted the strands become more tightly or more loosely wound. If the DNA is twisted in the direction of the helix, this is positive supercoiling, and the bases are held more tightly together. If they are twisted in the opposite direction, this is negative supercoiling, and the bases come apart more easily. In nature, most DNA has slight negative supercoiling that is introduced by enzymes called topoisomerases. These enzymes are also needed to relieve the twisting stresses introduced into DNA strands during processes such as transcription and DNA replication.

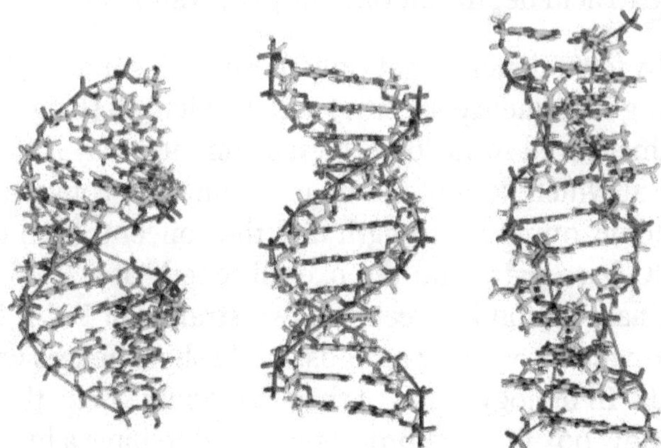

From left to right, the structures of A, B and Z DNA

Alternative DNA Structures

DNA exists in many possible conformations that include A-DNA, B-DNA, and Z-DNA forms, although, only B-DNA and Z-DNA have been directly observed in functional organisms. The conformation that DNA adopts depends on the hydration level, DNA sequence, the amount and direction of supercoiling, chemical modifications of the bases, the type and concentration of metal ions, and the presence of polyamines in solution.

The first published reports of A-DNA X-ray diffraction patterns—and also B-DNA—used analyses based on Patterson transforms that provided only a limited amount of structural information for oriented fibers of DNA. An alternative analysis was then proposed by Wilkins *et al.*, in 1953, for the *in vivo* B-DNA X-ray diffraction-scattering patterns of highly hydrated DNA fibers in terms of squares of Bessel functions. In the same journal, James Watson and Francis Crick presented their molecular modeling analysis of the DNA X-ray diffraction patterns to suggest that the structure was a double-helix.

Although the *B-DNA form* is most common under the conditions found in cells, it is not a well-defined conformation but a family of related DNA conformations that occur at the high hydration levels present in living cells. Their corresponding X-ray diffraction and scattering patterns are characteristic of molecular paracrystals with a significant degree of disorder.

Compared to B-DNA, the A-DNA form is a wider right-handed spiral, with a shallow, wide minor groove and a narrower, deeper major groove. The A form occurs under non-physiological conditions in partly dehydrated samples of DNA, while in the cell it may be produced in hybrid pairings of DNA and RNA strands, and in enzyme-DNA complexes. Segments of DNA where the bases have been chemically modified by methylation may undergo a larger change in conformation and adopt the Z form. Here, the strands turn about the helical axis in a left-handed spiral, the opposite of the more common B form. These unusual structures can be recognized by specific Z-DNA binding proteins and may be involved in the regulation of transcription.

Alternative DNA Chemistry

For many years exobiologists have proposed the existence of a shadow biosphere, a postulated microbial biosphere of Earth that uses radically different biochemical and molecular processes than currently known life. One of the proposals was the existence of lifeforms that use arsenic instead of phosphorus in DNA. A report in 2010 of the possibility in the bacterium GFAJ-1, was announced, though the research was disputed, and evidence suggests the bacterium actively prevents the incorporation of arsenic into the DNA backbone and other biomolecules.

Quadruplex Structures

At the ends of the linear chromosomes are specialized regions of DNA called telomeres. The main function of these regions is to allow the cell to replicate chromosome ends using the enzyme telomerase, as the enzymes that normally replicate DNA cannot copy the extreme 3′ ends of chromosomes. These specialized chromosome caps also help protect the DNA ends, and stop the DNA repair systems in the cell from treating them as damage to be corrected. In human cells, telomeres are usually lengths of single-stranded DNA containing several thousand repeats of a simple TTAGGG sequence.

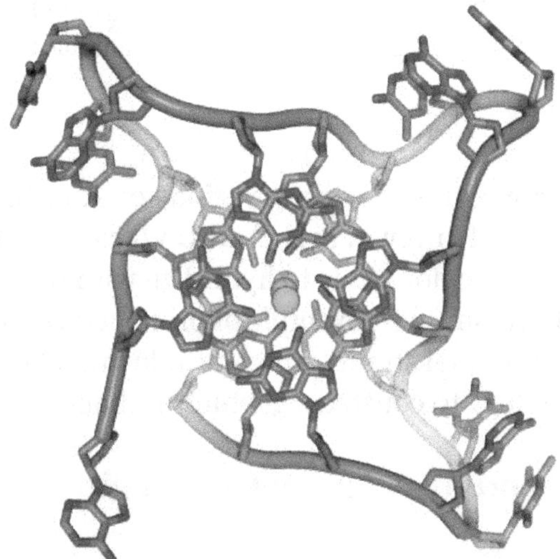

DNA quadruplex formed by telomere repeats. The looped conformation of the DNA backbone is very different from the typical DNA helix. The green spheres in the center represent potassium ions.

These guanine-rich sequences may stabilize chromosome ends by forming structures of stacked sets of four-base units, rather than the usual base pairs found in other DNA molecules. Here, four guanine bases form a flat plate and these flat four-base units then stack on top of each other, to form a stable G-quadruplex structure. These structures are stabilized by hydrogen bonding between the edges of the bases and chelation of a metal ion in the centre of each four-base unit. Other structures can also be formed, with the central set of four bases coming from either a single strand folded around the bases, or several different parallel strands, each contributing one base to the central structure.

In addition to these stacked structures, telomeres also form large loop structures called telomere loops, or T-loops. Here, the single-stranded DNA curls around in a long circle stabilized by telomere-binding proteins. At the very end of the T-loop, the single-stranded telomere DNA is held onto a region of double-stranded DNA by the telomere strand disrupting the double-helical DNA and base pairing to one of the two strands. This triple-stranded structure is called a displacement loop or D-loop.

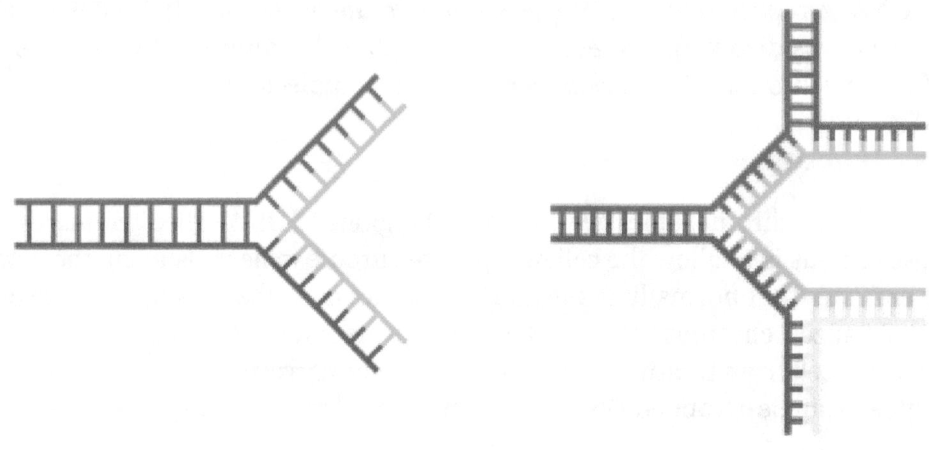

Single branch Multiple branches

Branched DNA can form networks containing multiple branches.

Branched DNA

In DNA, fraying occurs when non-complementary regions exist at the end of an otherwise complementary double-strand of DNA. However, branched DNA can occur if a third strand of DNA is introduced and contains adjoining regions able to hybridize with the frayed regions of the pre-existing double-strand. Although the simplest example of branched DNA involves only three strands of DNA, complexes involving additional strands and multiple branches are also possible. Branched DNA can be used in nanotechnology to construct geometric shapes.

Chemical Modi ications and Altered DNA Packaging

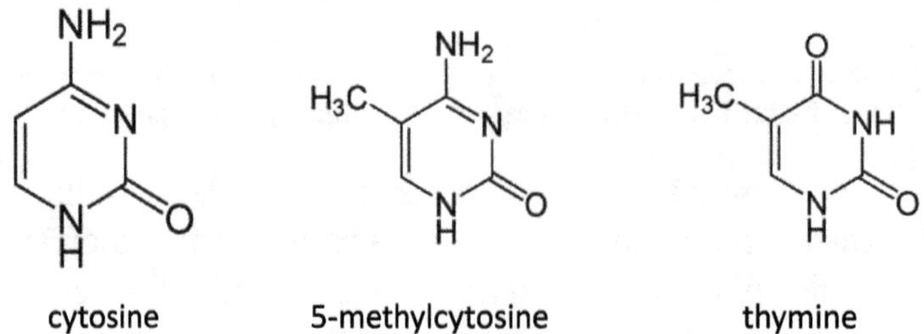

Structure of cytosine with and without the 5-methyl group. Deamination converts 5-methylcytosine into thymine.

Base Modifications and DNA Packaging

The expression of genes is influenced by how the DNA is packaged in chromosomes, in a structure called chromatin. Base modifications can be involved in packaging, with regions that have low or no gene expression usually containing high levels of methylation of cytosine bases. DNA packaging and its influence on gene expression can also occur by covalent modifications of the histone protein core around which DNA is wrapped in the chromatin structure or else by remodeling carried out by chromatin remodeling complexes. There is, further, crosstalk between DNA methylation and histone modification, so they can coordinately affect chromatin and gene expression.

For one example, cytosine methylation, produces 5-methylcytosine, which is important for X-inactivation of chromosomes. The average level of methylation varies between organisms – the worm *Caenorhabditis elegans* lacks cytosine methylation, while vertebrates have higher levels, with up to 1% of their DNA containing 5-methylcytosine. Despite the importance of 5-methylcytosine, it can deaminate to leave a thymine base, so methylated cytosines are particularly prone to mutations. Other base modifications include adenine methylation in bacteria, the presence of 5-hydroxymethylcytosine in the brain, and the glycosylation of uracil to produce the "J-base" in kinetoplastids.

Damage

DNA can be damaged by many sorts of mutagens, which change the DNA sequence. Mutagens include oxidizing agents, alkylating agents and also high-energy electromagnetic radiation such as ultraviolet light and X-rays. The type of DNA damage produced depends on the type of mutagen. For example, UV light can damage DNA by producing thymine dimers, which are cross-links between pyrimidine bases. On the other hand, oxidants such as free radicals or hydrogen peroxide produce multiple forms of damage, including base modifications, particularly of guanosine, and double-strand breaks. A typical human cell contains about 150,000 bases that have suffered oxidative damage. Of these oxidative lesions, the most dangerous are double-strand breaks, as these are difficult to repair and can produce point mutations, insertions, deletions from the DNA sequence, and chromosomal translocations. These mutations can cause cancer. Because of inherent limits in the DNA repair mechanisms, if humans lived long enough, they would all eventually develop cancer. DNA damages that are naturally occurring, due to normal cellular processes that produce reactive oxygen species, the hydrolytic activities of cellular water, etc., also occur frequently. Although most of these damages are repaired, in any cell some DNA damage may remain despite the action of repair processes. These remaining DNA damages accumulate with age in mammalian postmitotic tissues. This accumulation appears to be an important underlying cause of aging.

Many mutagens fit into the space between two adjacent base pairs, this is called *intercalation*. Most intercalators are aromatic and planar molecules; examples include ethidium bromide, acridines, daunomycin, and doxorubicin. For an intercalator to fit between base pairs, the bases must separate, distorting the DNA strands by unwinding of the double helix. This inhibits both transcription and DNA replication, causing toxicity and mutations. As a result, DNA intercalators may be carcinogens, and in the case of thalidomide, a teratogen. Others such as benzo[*a*]pyrene diol epoxide and aflatoxin form DNA adducts that induce errors in replication. Nevertheless, due to their ability to inhibit DNA transcription and replication, other similar toxins are also used in chemotherapy to inhibit rapidly growing cancer cells.

Biological Functions

DNA usually occurs as linear chromosomes in eukaryotes, and circular chromosomes in prokaryotes. The set of chromosomes in a cell makes up its genome; the human genome has approximately

3 billion base pairs of DNA arranged into 46 chromosomes. The information carried by DNA is held in the sequence of pieces of DNA called genes. Transmission of genetic information in genes is achieved via complementary base pairing. For example, in transcription, when a cell uses the information in a gene, the DNA sequence is copied into a complementary RNA sequence through the attraction between the DNA and the correct RNA nucleotides. Usually, this RNA copy is then used to make a matching protein sequence in a process called translation, which depends on the same interaction between RNA nucleotides. In alternative fashion, a cell may simply copy its genetic information in a process called DNA replication.

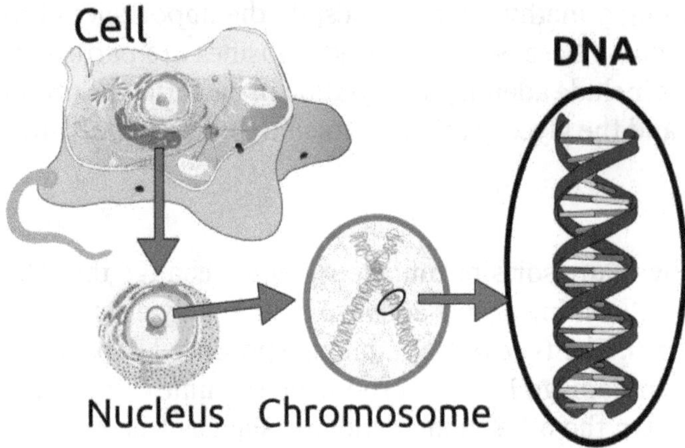

Location of eukaryote nuclear DNA within the chromosomes.

Genes and Genomes

Genomic DNA is tightly and orderly packed in the process called DNA condensation, to fit the small available volumes of the cell. In eukaryotes, DNA is located in the cell nucleus, with small amounts in mitochondria and chloroplasts. In prokaryotes, the DNA is held within an irregularly shaped body in the cytoplasm called the nucleoid. The genetic information in a genome is held within genes, and the complete set of this information in an organism is called its genotype. A gene is a unit of heredity and is a region of DNA that influences a particular characteristic in an organism. Genes contain an open reading frame that can be transcribed, and regulatory sequences such as promoters and enhancers, which control transcription of the open reading frame.

In many species, only a small fraction of the total sequence of the genome encodes protein. For example, only about 1.5% of the human genome consists of protein-coding exons, with over 50% of human DNA consisting of non-coding repetitive sequences. The reasons for the presence of so much noncoding DNA in eukaryotic genomes and the extraordinary differences in genome size, or *C-value*, among species represent a long-standing puzzle known as the "C-value enigma". However, some DNA sequences that do not code protein may still encode functional non-coding RNA molecules, which are involved in the regulation of gene expression.

Some noncoding DNA sequences play structural roles in chromosomes. Telomeres and centromeres typically contain few genes, but are important for the function and stability of chromosomes. An abundant form of noncoding DNA in humans are pseudogenes, which are copies of genes that have been disabled by mutation. These sequences are usually just molecular fossils, although they can occasionally serve as raw genetic material for the creation of new genes through the process of gene duplication and divergence.

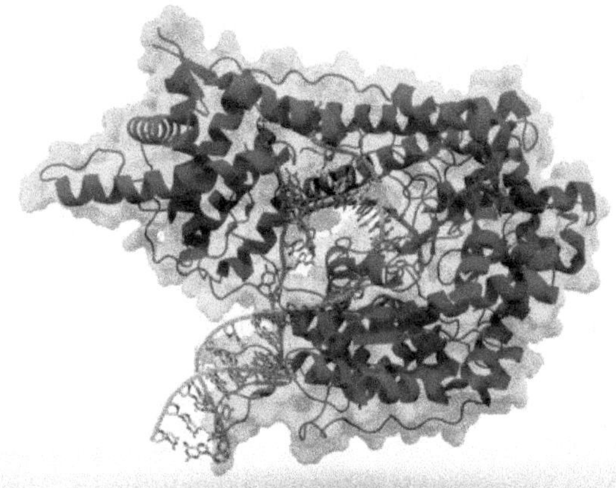

T7 RNA polymerase (blue) producing a mRNA (green) from a DNA template (orange).

Transcription and Translation

A gene is a sequence of DNA that contains genetic information and can influence the phenotype of an organism. Within a gene, the sequence of bases along a DNA strand defines a messenger RNA sequence, which then defines one or more protein sequences. The relationship between the nucleotide sequences of genes and the amino-acid sequences of proteins is determined by the rules of translation, known collectively as the genetic code. The genetic code consists of three-letter 'words' called *codons* formed from a sequence of three nucleotides (e.g. ACT, CAG, TTT).

In transcription, the codons of a gene are copied into messenger RNA by RNA polymerase. This RNA copy is then decoded by a ribosome that reads the RNA sequence by base-pairing the messenger RNA to transfer RNA, which carries amino acids. Since there are 4 bases in 3-letter combinations, there are 64 possible codons (4^3 combinations). These encode the twenty standard amino acids, giving most amino acids more than one possible codon. There are also three 'stop' or 'nonsense' codons signifying the end of the coding region; these are the TAA, TGA, and TAG codons.

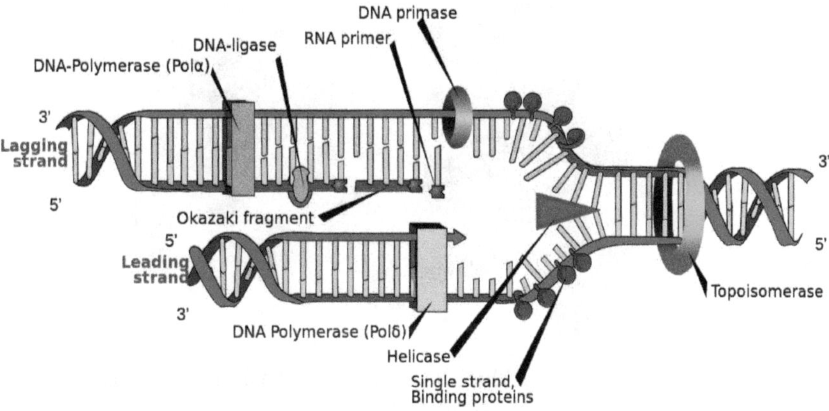

DNA replication. The double helix is unwound by a helicase and topoisomerase. Next, one DNA polymerase produces the leading strand copy. Another DNA polymerase binds to the lagging strand. This enzyme makes discontinuous segments (called Okazaki fragments) before DNA ligase joins them together.

Replication

Cell division is essential for an organism to grow, but, when a cell divides, it must replicate the DNA in its genome so that the two daughter cells have the same genetic information as their parent. The double-stranded structure of DNA provides a simple mechanism for DNA replication. Here, the two strands are separated and then each strand's complementary DNA sequence is recreated by an enzyme called DNA polymerase. This enzyme makes the complementary strand by finding the correct base through complementary base pairing, and bonding it onto the original strand. As DNA polymerases can only extend a DNA strand in a 5′ to 3′ direction, different mechanisms are used to copy the antiparallel strands of the double helix. In this way, the base on the old strand dictates which base appears on the new strand, and the cell ends up with a perfect copy of its DNA.

Extracellular Nucleic Acids

Naked extracellular DNA (eDNA), most of it released by cell death, is nearly ubiquitous in the environment. Its concentration in soil may be as high as 2 μg/L, and its concentration in natural aquatic environments may be as high at 88 μg/L. Various possible functions have been proposed for eDNA: it may be involved in horizontal gene transfer; it may provide nutrients; and it may act as a buffer to recruit or titrate ions or antibiotics. Extracellular DNA acts as a functional extracellular matrix component in the biofilms of several bacterial species. It may act as a recognition factor to regulate the attachment and dispersal of specific cell types in the biofilm; it may contribute to biofilm formation; and it may contribute to the biofilm's physical strength and resistance to biological stress.

Interactions with Proteins

All the functions of DNA depend on interactions with proteins. These protein interactions can be non-specific, or the protein can bind specifically to a single DNA sequence. Enzymes can also bind to DNA and of these, the polymerases that copy the DNA base sequence in transcription and DNA replication are particularly important.

DNA-binding Proteins

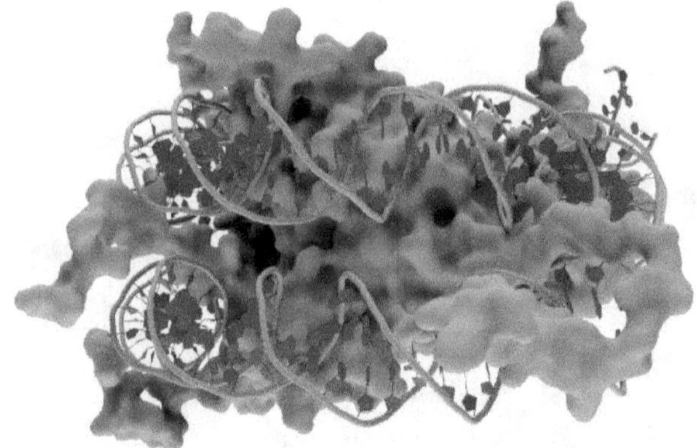

Interaction of DNA (in orange) with histones (in blue). These proteins' basic amino acids bind to the acidic phosphate groups on DNA.

Structural proteins that bind DNA are well-understood examples of non-specific DNA-protein interactions. Within chromosomes, DNA is held in complexes with structural proteins. These proteins organize the DNA into a compact structure called chromatin. In eukaryotes this structure involves DNA binding to a complex of small basic proteins called histones, while in prokaryotes multiple types of proteins are involved. The histones form a disk-shaped complex called a nucleosome, which contains two complete turns of double-stranded DNA wrapped around its surface. These non-specific interactions are formed through basic residues in the histones, making ionic bonds to the acidic sugar-phosphate backbone of the DNA, and are thus largely independent of the base sequence. Chemical modifications of these basic amino acid residues include methylation, phosphorylation and acetylation. These chemical changes alter the strength of the interaction between the DNA and the histones, making the DNA more or less accessible to transcription factors and changing the rate of transcription. Other non-specific DNA-binding proteins in chromatin include the high-mobility group proteins, which bind to bent or distorted DNA. These proteins are important in bending arrays of nucleosomes and arranging them into the larger structures that make up chromosomes.

A distinct group of DNA-binding proteins are the DNA-binding proteins that specifically bind single-stranded DNA. In humans, replication protein A is the best-understood member of this family and is used in processes where the double helix is separated, including DNA replication, recombination and DNA repair. These binding proteins seem to stabilize single-stranded DNA and protect it from forming stem-loops or being degraded by nucleases.

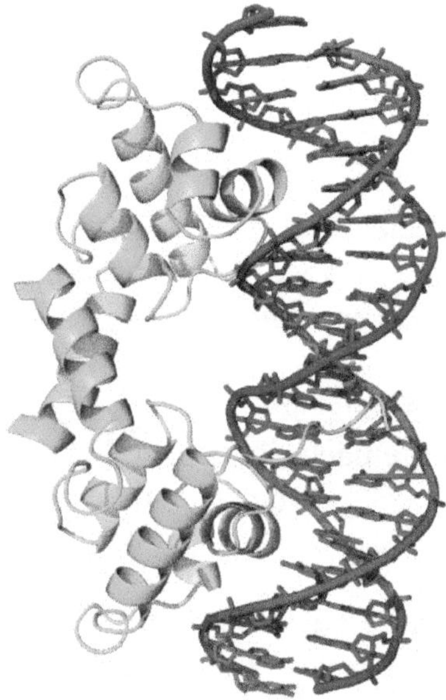

The lambda repressor helix-turn-helix transcription factor bound to its DNA target

In contrast, other proteins have evolved to bind to particular DNA sequences. The most intensively studied of these are the various transcription factors, which are proteins that regulate transcription. Each transcription factor binds to one particular set of DNA sequences and activates or inhibits the transcription of genes that have these sequences close to their promoters. The transcription factors do this in two ways. Firstly, they can bind the RNA polymerase responsible for transcription, either directly or through other mediator proteins; this locates the polymerase at the promoter and allows it to begin transcription. Alternatively, transcription factors can bind enzymes that modify the histones at the promoter. This changes the accessibility of the DNA template to the polymerase.

As these DNA targets can occur throughout an organism's genome, changes in the activity of one type of transcription factor can affect thousands of genes. Consequently, these proteins are often the targets of the signal transduction processes that control responses to environmental changes or cellular differentiation and development. The specificity of these transcription factors' interactions with DNA come from the proteins making multiple contacts to the edges of the DNA bases, allowing them to "read" the DNA sequence. Most of these base-interactions are made in the major groove, where the bases are most accessible.

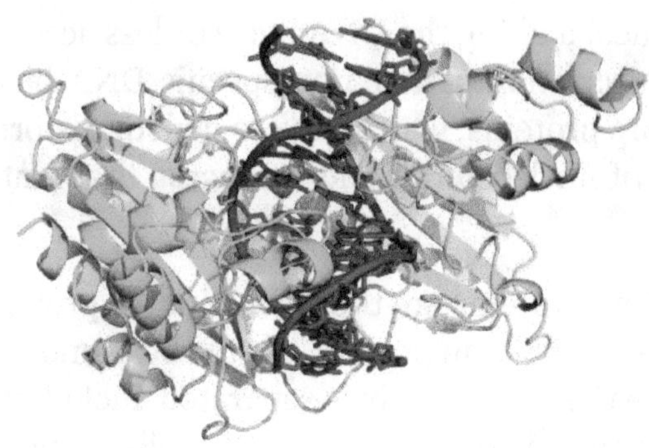

The restriction enzyme EcoRV (green) in a complex with its substrate DNA

DNA-modifying Enzymes

Nucleases and Ligases

Nucleases are enzymes that cut DNA strands by catalyzing the hydrolysis of the phosphodiester bonds. Nucleases that hydrolyse nucleotides from the ends of DNA strands are called exonucleases, while endonucleases cut within strands. The most frequently used nucleases in molecular biology are the restriction endonucleases, which cut DNA at specific sequences. For instance, the EcoRV enzyme shown to the left recognizes the 6-base sequence 5′-GATATC-3′ and makes a cut at the horizontal line. In nature, these enzymes protect bacteria against phage infection by digesting the phage DNA when it enters the bacterial cell, acting as part of the restriction modification system. In technology, these sequence-specific nucleases are used in molecular cloning and DNA fingerprinting.

Enzymes called DNA ligases can rejoin cut or broken DNA strands. Ligases are particularly important in lagging strand DNA replication, as they join together the short segments of DNA produced at the replication fork into a complete copy of the DNA template. They are also used in DNA repair and genetic recombination.

Topoisomerases and Helicases

Topoisomerases are enzymes with both nuclease and ligase activity. These proteins change the amount of supercoiling in DNA. Some of these enzymes work by cutting the DNA helix and allowing one section to rotate, thereby reducing its level of supercoiling; the enzyme then seals the DNA break. Other types of these enzymes are capable of cutting one DNA helix and then passing a second strand of DNA through this break, before rejoining the helix. Topoisomerases are required for many processes involving DNA, such as DNA replication and transcription.

Helicases are proteins that are a type of molecular motor. They use the chemical energy in nucleoside triphosphates, predominantly adenosine triphosphate (ATP), to break hydrogen bonds between bases and unwind the DNA double helix into single strands. These enzymes are essential for most processes where enzymes need to access the DNA bases.

Polymerases

Polymerases are enzymes that synthesize polynucleotide chains from nucleoside triphosphates. The sequence of their products are created based on existing polynucleotide chains—which are called *templates*. These enzymes function by repeatedly adding a nucleotide to the 3′ hydroxyl group at the end of the growing polynucleotide chain. As a consequence, all polymerases work in a 5′ to 3′ direction. In the active site of these enzymes, the incoming nucleoside triphosphate base-pairs to the template: this allows polymerases to accurately synthesize the complementary strand of their template. Polymerases are classified according to the type of template that they use.

In DNA replication, DNA-dependent DNA polymerases make copies of DNA polynucleotide chains. To preserve biological information, it is essential that the sequence of bases in each copy are precisely complementary to the sequence of bases in the template strand. Many DNA polymerases have a proofreading activity. Here, the polymerase recognizes the occasional mistakes in the synthesis reaction by the lack of base pairing between the mismatched nucleotides. If a mismatch is detected, a 3′ to 5′ exonuclease activity is activated and the incorrect base removed. In most organisms, DNA polymerases function in a large complex called the replisome that contains multiple accessory subunits, such as the DNA clamp or helicases.

RNA-dependent DNA polymerases are a specialized class of polymerases that copy the sequence of an RNA strand into DNA. They include reverse transcriptase, which is a viral enzyme involved in the infection of cells by retroviruses, and telomerase, which is required for the replication of telomeres. Telomerase is an unusual polymerase because it contains its own RNA template as part of its structure.

Transcription is carried out by a DNA-dependent RNA polymerase that copies the sequence of a DNA strand into RNA. To begin transcribing a gene, the RNA polymerase binds to a sequence of DNA called a promoter and separates the DNA strands. It then copies the gene sequence into a messenger RNA transcript until it reaches a region of DNA called the terminator, where it halts and detaches from the DNA. As with human DNA-dependent DNA polymerases, RNA polymerase II, the enzyme that transcribes most of the genes in the human genome, operates as part of a large protein complex with multiple regulatory and accessory subunits.

Genetic Recombination

A DNA helix usually does not interact with other segments of DNA, and in human cells the different chromosomes even occupy separate areas in the nucleus called “chromosome territories”. This physical separation of different chromosomes is important for the ability of DNA to function as a stable repository for information, as one of the few times chromosomes interact is in chromosomal crossover which occurs during sexual reproduction, when genetic recombination occurs. Chromosomal crossover is when two DNA helices break, swap a section and then rejoin.

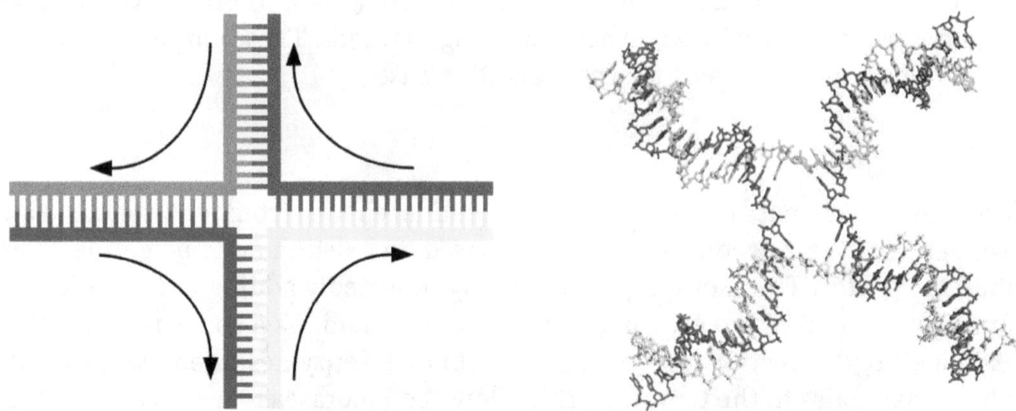

Structure of the Holliday junction intermediate in genetic recombination. The four separate DNA strands are coloured red, blue, green and yellow.

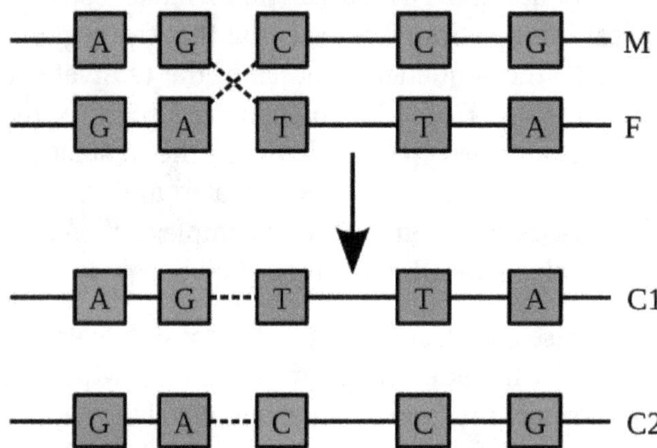

Recombination involves the breaking and rejoining of two chromosomes (M and F) to produce two rearranged chromosomes (C1 and C2).

Recombination allows chromosomes to exchange genetic information and produces new combinations of genes, which increases the efficiency of natural selection and can be important in the rapid evolution of new proteins. Genetic recombination can also be involved in DNA repair, particularly in the cell's response to double-strand breaks.

The most common form of chromosomal crossover is homologous recombination, where the two chromosomes involved share very similar sequences. Non-homologous recombination can be damaging to cells, as it can produce chromosomal translocations and genetic abnormalities. The recombination reaction is catalyzed by enzymes known as recombinases, such as RAD51. The first step in recombination is a double-stranded break caused by either an endonuclease or damage to the DNA. A series of steps catalyzed in part by the recombinase then leads to joining of the two helices by at least one Holliday junction, in which a segment of a single strand in each helix is annealed to the complementary strand in the other helix. The Holliday junction is a tetrahedral junction structure that can be moved along the pair of chromosomes, swapping one strand for another. The recombination reaction is then halted by cleavage of the junction and re-ligation of the released DNA.

Evolution

DNA contains the genetic information that allows all modern living things to function, grow and reproduce. However, it is unclear how long in the 4-billion-year history of life DNA has performed this function, as it has been proposed that the earliest forms of life may have used RNA as their genetic material. RNA may have acted as the central part of early cell metabolism as it can both transmit genetic information and carry out catalysis as part of ribozymes. This ancient RNA world where nucleic acid would have been used for both catalysis and genetics may have influenced the evolution of the current genetic code based on four nucleotide bases. This would occur, since the number of different bases in such an organism is a trade-off between a small number of bases increasing replication accuracy and a large number of bases increasing the catalytic efficiency of ribozymes. However, there is no direct evidence of ancient genetic systems, as recovery of DNA from most fossils is impossible because DNA survives in the environment for less than one million years, and slowly degrades into short fragments in solution. Claims for older DNA have been made, most notably a report of the isolation of a viable bacterium from a salt crystal 250 million years old, but these claims are controversial.

Building blocks of DNA (adenine, guanine and related organic molecules) may have been formed extraterrestrially in outer space. Complex DNA and RNA organic compounds of life, including uracil, cytosine, and thymine, have also been formed in the laboratory under conditions mimicking those found in outer space, using starting chemicals, such as pyrimidine, found in meteorites. Pyrimidine, like polycyclic aromatic hydrocarbons (PAHs), the most carbon-rich chemical found in the universe, may have been formed in red giants or in interstellar cosmic dust and gas clouds.

Uses in Technology

Genetic Engineering

Methods have been developed to purify DNA from organisms, such as phenol-chloroform extraction, and to manipulate it in the laboratory, such as restriction digests and the polymerase chain reaction. Modern biology and biochemistry make intensive use of these techniques in recombinant DNA technology. Recombinant DNA is a man-made DNA sequence that has been assembled from other DNA sequences. They can be transformed into organisms in the form of plasmids or in the appropriate format, by using a viral vector. The genetically modified organisms produced can be used to produce products such as recombinant proteins, used in medical research, or be grown in agriculture.

DNA Profiling

Forensic scientists can use DNA in blood, semen, skin, saliva or hair found at a crime scene to identify a matching DNA of an individual, such as a perpetrator. This process is formally termed DNA profiling, but may also be called "genetic fingerprinting". In DNA profiling, the lengths of variable sections of repetitive DNA, such as short tandem repeats and minisatellites, are compared between people. This method is usually an extremely reliable technique for identifying a matching DNA. However, identification can be complicated if the scene is contaminated with DNA from several people. DNA profiling was developed in 1984 by British geneticist Sir Alec Jeffreys, and first used in forensic science to convict Colin Pitchfork in the 1988 Enderby murders case.

The development of forensic science, and the ability to now obtain genetic matching on minute samples of blood, skin, saliva, or hair has led to re-examining many cases. Evidence can now be uncovered that was scientifically impossible at the time of the original examination. Combined with the removal of the double jeopardy law in some places, this can allow cases to be reopened where prior trials have failed to produce sufficient evidence to convince a jury. People charged with serious crimes may be required to provide a sample of DNA for matching purposes. The most obvious defence to DNA matches obtained forensically is to claim that cross-contamination of evidence has occurred. This has resulted in meticulous strict handling procedures with new cases of serious crime. DNA profiling is also used successfully to positively identify victims of mass casualty incidents, bodies or body parts in serious accidents, and individual victims in mass war graves, via matching to family members.

DNA profiling is also used in DNA paternity testing to determine if someone is the biological parent or grandparent of a child with the probability of parentage is typically 99.99% when the alleged parent is biologically related to the child. Normal DNA sequencing methods happen after birth but there are new methods to test paternity while a mother is still pregnant.

DNA Enzymes or Catalytic DNA

Deoxyribozymes, also called DNAzymes or catalytic DNA are first discovered in 1994. They are mostly single stranded DNA sequences isolated from a large pool of random DNA sequences through a combinatorial approach called in vitro selection or systematic evolution of ligands by exponential enrichment (SELEX). DNAzymes catalyze variety of chemical reactions including RNA-DNA cleavage, RNA-DNA ligation, amino acids phosphorylation-dephosphorylation, carbon-carbon bond formation, and etc. DNAzymes can enhance catalytic rate of chemical reactions up to 100,000,000,000-fold over the uncatalyzed reaction. The most extensively studied class of DNAzymes are RNA-cleaving types which have been used to detect different metal ions and designing therapeutic agents. Several metal-specific DNAzymes have been reported including the GR-5 DNAzyme (lead-specific), the CA1-3 DNAzymes (copper-specific), the 39E DNAzyme (uranyl-specific) and the NaA43 DNAzyme (sodium-specific). The NaA43 DNAzyme, which is reported to be more than 10,000-fold selective for sodium over other metal ions, was used to make a real-time sodium sensor in living cells.

Bioinformatics

Bioinformatics involves the development of techniques to store, data mine, search and manipulate biological data, including DNA nucleic acid sequence data. These have led to widely applied advances in computer science, especially string searching algorithms, machine learning and database theory. String searching or matching algorithms, which find an occurrence of a sequence of letters inside a larger sequence of letters, were developed to search for specific sequences of nucleotides. The DNA sequence may be aligned with other DNA sequences to identify homologous sequences and locate the specific mutations that make them distinct. These techniques, especially multiple sequence alignment, are used in studying phylogenetic relationships and protein function. Data sets representing entire genomes' worth of DNA sequences, such as those produced by the Human Genome Project, are difficult to use without the annotations that identify the locations of genes and regulatory elements on each chromosome. Regions of DNA sequence that have the characteristic patterns associated with protein- or RNA-coding genes can be identified by gene finding algorithms, which allow researchers to predict the presence of particular gene products and their possible functions in an organism even before they have been isolated experimentally. Entire genomes may also be compared, which can shed light on the evolutionary history of particular organism and permit the examination of complex evolutionary events.

DNA Nanotechnology

DNA nanotechnology uses the unique molecular recognition properties of DNA and other nucleic acids to create self-assembling branched DNA complexes with useful properties. DNA is thus used as a structural material rather than as a carrier of biological information. This has led to the creation of two-dimensional periodic lattices (both tile-based and using the *DNA origami* method) and three-dimensional structures in the shapes of polyhedra. Nanomechanical devices and algorithmic self-assembly have also been demonstrated, and these DNA structures have been used to template the arrangement of other molecules such as gold nanoparticles and streptavidin proteins.

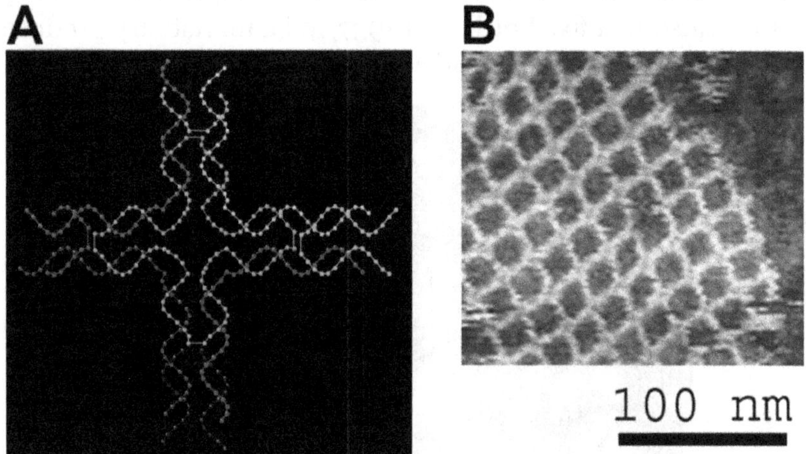

The DNA structure at left (schematic shown) will self-assemble into the structure visualized by atomic force microscopy at right. DNA nanotechnology is the field that seeks to design nanoscale structures using the molecular recognition properties of DNA molecules. Image from Strong, 2004.

History and Anthropology

Because DNA collects mutations over time, which are then inherited, it contains historical information, and, by comparing DNA sequences, geneticists can infer the evolutionary history of organisms, their phylogeny. This field of phylogenetics is a powerful tool in evolutionary biology. If DNA sequences within a species are compared, population geneticists can learn the history of particular populations. This can be used in studies ranging from ecological genetics to anthropology; For example, DNA evidence is being used to try to identify the Ten Lost Tribes of Israel.

Information Storage

In a paper published in *Nature* in January 2013, scientists from the European Bioinformatics Institute and Agilent Technologies proposed a mechanism to use DNA's ability to code information as a means of digital data storage. The group was able to encode 739 kilobytes of data into DNA code, synthesize the actual DNA, then sequence the DNA and decode the information back to its original form, with a reported 100% accuracy. The encoded information consisted of text files and audio files. A prior experiment was published in August 2012. It was conducted by researchers at Harvard University, where the text of a 54,000-word book was encoded in DNA.

Moreover, in living cells the storage can be turned active by enzymes. Light-gated protein domains fused to DNA processing enzymes are suitable for that task *in vitro*. Fluorescent exonucleases can transmit the output according to the nucleotide they have read.

History of DNA Research

DNA was first isolated by the Swiss physician Friedrich Miescher who, in 1869, discovered a microscopic substance in the pus of discarded surgical bandages. As it resided in the nuclei of cells, he called it "nuclein". In 1878, Albrecht Kossel isolated the non-protein component of "nuclein", nucleic acid, and later isolated its five primary nucleobases. In 1919, Phoebus Levene identified the base, sugar and phosphate nucleotide unit. Levene suggested that DNA consisted of a string of nucleotide units linked together through the phosphate groups. Levene thought the chain was short and the bases repeated in a fixed order. In 1937, William Astbury produced the first X-ray diffraction patterns that showed that DNA had a regular structure.

James Watson and Francis Crick (right), co-originators of the double-helix model, with Maclyn McCarty (left).

Pencil sketch of the DNA double helix by Francis Crick in 1953

In 1927, Nikolai Koltsov proposed that inherited traits would be inherited via a "giant hereditary molecule" made up of "two mirror strands that would replicate in a semi-conservative fashion using each strand as a template". In 1928, Frederick Griffith in his experiment discovered that traits of the "smooth" form of *Pneumococcus* could be transferred to the "rough" form of the same bacteria by mixing killed "smooth" bacteria with the live "rough" form. This system provided the first clear suggestion that DNA carries genetic information—the Avery–MacLeod–McCarty experiment—when Oswald Avery, along with coworkers Colin MacLeod and Maclyn McCarty, identified DNA as the transforming principle in 1943. DNA's role in heredity was confirmed in 1952, when Alfred Hershey and Martha Chase in the Hershey–Chase experiment showed that DNA is the genetic material of the T2 phage.

In 1953, James Watson and Francis Crick suggested what is now accepted as the first correct double-helix model of DNA structure in the journal *Nature*. Their double-helix, molecular model of DNA was then based on one X-ray diffraction image (labeled as "Photo 51") taken by Rosalind Franklin and Raymond Gosling in May 1952, and the information that the DNA bases are paired.

Experimental evidence supporting the Watson and Crick model was published in a series of five articles in the same issue of *Nature*. Of these, Franklin and Gosling's paper was the first publication of their own X-ray diffraction data and original analysis method that partly supported the Watson and Crick model; this issue also contained an article on DNA structure by Maurice Wilkins and two of his colleagues, whose analysis and *in vivo* B-DNA X-ray patterns also supported the presence *in vivo* of the double-helical DNA configurations as proposed by Crick and Watson for their double-helix molecular model of DNA in the prior two pages of *Nature*. In 1962, after Franklin's death, Watson, Crick, and Wilkins jointly received the Nobel Prize in Physiology or Medicine. Nobel Prizes are awarded only to living recipients. A debate continues about who should receive credit for the discovery.

In an influential presentation in 1957, Crick laid out the central dogma of molecular biology, which foretold the relationship between DNA, RNA, and proteins, and articulated the "adaptor hypothesis". Final confirmation of the replication mechanism that was implied by the double-helical structure followed in 1958 through the Meselson–Stahl experiment. Further work by Crick and coworkers showed that the genetic code was based on non-overlapping triplets of bases, called codons, allowing Har Gobind Khorana, Robert W. Holley and Marshall Warren Nirenberg to decipher the genetic code. These findings represent the birth of molecular biology.

Noncoding DNA

In genomics and related disciplines, noncoding DNA sequences are components of an organism's DNA that do not encode protein sequences. Some noncoding DNA is transcribed into functional non-coding RNA molecules (e.g. transfer RNA, ribosomal RNA, and regulatory RNAs). Other functions of noncoding DNA include the transcriptional and translational regulation of protein-coding sequences, scaffold attachment regions, origins of DNA replication, centromeres and telomeres.

The amount of noncoding DNA varies greatly among species. Where only a small percentage of the genome is responsible for coding proteins, the percentage of the genome performing regulatory functions is growing. When there is much non-coding DNA, a large proportion appears to have no biological function for the organism, as theoretically predicted in the 1960s. Since that time, this non-functional portion has often been referred to as "junk DNA", a term that has elicited strong responses over the years.

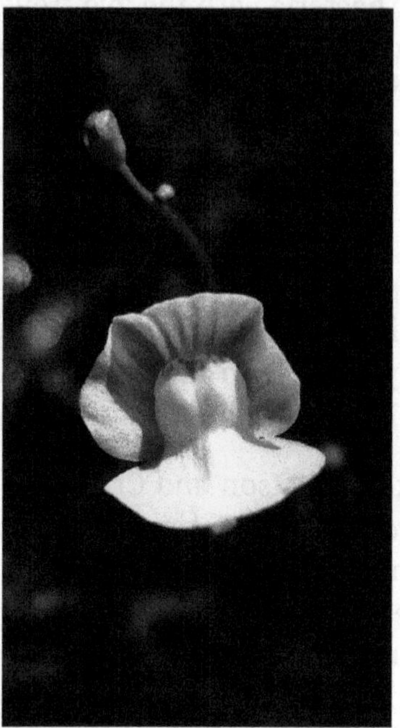

Utricularia gibba has only 3% noncoding DNA.

The international Encyclopedia of DNA Elements (ENCODE) project uncovered, by direct biochemical approaches, that at least 80% of human genomic DNA has biochemical activity. Though this was not necessarily unexpected due to previous decades of research discovering many functional noncoding regions, some scientists criticized the conclusion for conflating biochemical activity with biological function. Estimates for the biologically functional fraction of our genome based on comparative genomics range between 8 and 15%. However, others have argued against relying solely on estimates from comparative genomics due to its limited scope because non-coding DNA has been found to be involved in epigenetic activity and complex networks of genetic interactions, being explored in evolutionary developmental biology.

Fraction of Noncoding Genomic DNA

The amount of total genomic DNA varies widely between organisms, and the proportion of coding and noncoding DNA within these genomes varies greatly as well. For example, it was originally suggested that over 98% of the human genome does not encode protein sequences, including most sequences within introns and most intergenic DNA, whilst 20% of a typical prokaryote genome is noncoding.

While overall genome size, and by extension the amount of noncoding DNA, are correlated to organism complexity, there are many exceptions. For example, the genome of the unicellular *Polychaos dubium* (formerly known as *Amoeba dubia*) has been reported to contain more than 200 times the amount of DNA in humans. The pufferfish *Takifugu rubripes* genome is only about one eighth the size of the human genome, yet seems to have a comparable number of genes; approximately 90%

of the *Takifugu* genome is noncoding DNA. The extensive variation in nuclear genome size among eukaryotic species is known as the C-value enigma or C-value paradox. Most of the genome size difference appears to lie in the noncoding DNA.

In 2013, a new "record" for the most efficient eukaryotic genome was discovered with *Utricularia gibba*, a bladderwort plant that has only 3% noncoding DNA and 97% of coding DNA. Parts of the noncoding DNA were being deleted by the plant and this suggested that noncoding DNA may not be as critical for plants, even though noncoding DNA is useful for humans. Other studies on plants have discovered crucial functions in portions noncoding DNA that were previously thought to be negligible and have added a new layer to the understanding of gene regulation.

Types of Noncoding DNA Sequences

Noncoding Functional RNA

Noncoding RNAs are functional RNA molecules that are not translated into protein. Examples of noncoding RNA include ribosomal RNA, transfer RNA, Piwi-interacting RNA and microRNA.

MicroRNAs are predicted to control the translational activity of approximately 30% of all protein-coding genes in mammals and may be vital components in the progression or treatment of various diseases including cancer, cardiovascular disease, and the immune system response to infection.

Cis- and Trans-regulatory Elements

Cis-regulatory elements are sequences that control the transcription of a nearby gene. Many such elements are involved in the evolution and control of development. Cis-elements may be located in 5' or 3' untranslated regions or within introns. Trans-regulatory elements control the transcription of a distant gene.

Promoters facilitate the transcription of a particular gene and are typically upstream of the coding region. Enhancer sequences may also exert very distant effects on the transcription levels of genes.

Introns

Introns are non-coding sections of a gene, transcribed into the precursor mRNA sequence, but ultimately removed by RNA splicing during the processing to mature messenger RNA. Many introns appear to be mobile genetic elements.

Studies of group I introns from *Tetrahymena* protozoans indicate that some introns appear to be selfish genetic elements, neutral to the host because they remove themselves from flanking exons during RNA processing and do not produce an expression bias between alleles with and without the intron. Some introns appear to have significant biological function, possibly through ribozyme functionality that may regulate tRNA and rRNA activity as well as protein-coding gene expression, evident in hosts that have become dependent on such introns over long periods of time; for example, the *trnL-intron* is found in all green plants and appears to have been vertically inherited for several billions of years, including more than a billion years within chloroplasts and an additional 2–3 billion years prior in the cyanobacterial ancestors of chloroplasts.

Pseudogenes

Pseudogenes are DNA sequences, related to known genes, that have lost their protein-coding ability or are otherwise no longer expressed in the cell. Pseudogenes arise from retrotransposition or genomic duplication of functional genes, and become "genomic fossils" that are nonfunctional due to mutations that prevent the transcription of the gene, such as within the gene promoter region, or fatally alter the translation of the gene, such as premature stop codons or frameshifts. Pseudogenes resulting from the retrotransposition of an RNA intermediate are known as processed pseudogenes; pseudogenes that arise from the genomic remains of duplicated genes or residues of inactivated genes are nonprocessed pseudogenes.

While Dollo's Law suggests that the loss of function in pseudogenes is likely permanent, silenced genes may actually retain function for several million years and can be "reactivated" into protein-coding sequences and a substantial number of pseudogenes are actively transcribed. Because pseudogenes are presumed to change without evolutionary constraint, they can serve as a useful model of the type and frequencies of various spontaneous genetic mutations.

Repeat Sequences, Transposons and Viral Elements

Transposons and retrotransposons are mobile genetic elements. Retrotransposon repeated sequences, which include long interspersed nuclear elements (LINEs) and short interspersed nuclear elements (SINEs), account for a large proportion of the genomic sequences in many species. Alu sequences, classified as a short interspersed nuclear element, are the most abundant mobile elements in the human genome. Some examples have been found of SINEs exerting transcriptional control of some protein-encoding genes.

Endogenous retrovirus sequences are the product of reverse transcription of retrovirus genomes into the genomes of germ cells. Mutation within these retro-transcribed sequences can inactivate the viral genome.

Over 8% of the human genome is made up of (mostly decayed) endogenous retrovirus sequences, as part of the over 42% fraction that is recognizably derived of retrotransposons, while another 3% can be identified to be the remains of DNA transposons. Much of the remaining half of the genome that is currently without an explained origin is expected to have found its origin in transposable elements that were active so long ago (> 200 million years) that random mutations have rendered them unrecognizable. Genome size variation in at least two kinds of plants is mostly the result of retrotransposon sequences.

Telomeres

Telomeres are regions of repetitive DNA at the end of a chromosome, which provide protection from chromosomal deterioration during DNA replication.

Junk DNA

The term "junk DNA" became popular in the 1960s. According to T. Ryan Gregory, a genomic biologist, David Comings was the first to discuss the nature of junk DNA explicitly in 1972, and he applied the term to all noncoding DNA. The term was formalized in 1972 by Susumu Ohno, who noted that the mutational load from deleterious mutations placed an upper limit on the number of functional loci that could be expected given a typical mutation rate. Ohno hypothesized that mammal genomes could not have more than 30,000 loci under selection before the "cost" from the mutational load would cause an inescapable decline in fitness, and eventually extinction. This prediction remains robust, with the human genome containing approximately 20,000 genes. Another source for Ohno's theory was the observation that even closely related species can have widely (orders-of-magnitude) different genome sizes, which had been dubbed the C-value paradox in 1971.

Though the fruitfulness of the term "junk DNA" has been questioned on the grounds that it provokes a strong a priori assumption of total non-functionality and though some have recommended using more neutral terminology such as "noncoding DNA" instead; "junk DNA" remains a label for the portions of a genome sequence for which no discernible function has been identified and that through comparative genomics analysis appear under no functional constraint suggesting that the sequence itself has provided no adaptive advantage. Since the late 70s it has become apparent that the majority of non-coding DNA in large genomes finds its origin in the selfish amplification of transposable elements, of which W. Ford Doolittle and Carmen Sapienza in 1980 wrote in the journal *Nature*: "When a given DNA, or class of DNAs, of unproven phenotypic function can be shown to have evolved a strategy (such as transposition) which ensures its genomic survival, then no other explanation for its existence is necessary." The amount of junk DNA can be expected to depend on the rate of amplification of these elements and the rate at which non-functional DNA is lost. In the same issue of *Nature*, Leslie Orgel and Francis Crick wrote that junk DNA has "little specificity and conveys little or no selective advantage to the organism". The term occurs mainly in popular science and in a colloquial way in scientific publications, and it has been suggested that its connotations may have delayed interest in the biological functions of noncoding DNA.

Several lines of evidence indicate that some "junk DNA" sequences are likely to have unidentified functional activity and that the process of exaptation of fragments of originally selfish or non-functional DNA has been commonplace throughout evolution. In 2012, the ENCODE project, a research program supported by the National Human Genome Research Institute, reported that 76% of the human genome's noncoding DNA sequences were transcribed and that nearly half of the genome was in some way accessible to genetic regulatory proteins such as transcription factors.

However, the suggestion by ENCODE that over 80% of the human genome is biochemically functional has been criticized by other scientists, who argue that neither accessibility of segments of the genome to transcription factors nor their transcription guarantees that those segments have biochemical function and that their transcription is selectively advantageous. Furthermore, the much lower estimates of functionality prior to ENCODE were based on genomic conservation estimates across mammalian lineages.

In response to such views, other scientists argue that the wide spread transcription and splicing that is observed in the human genome directly by biochemical testing is a more accurate indicator of genetic function than genomic conservation because conservation estimates are relative due to incredible variations in genome sizes of even closely related species, it is partially tautological, and these estimates are not based on direct testing for functionality on the genome. Conservation estimates may be used to provide clues to identify possible functional elements in the genome, but it does not limit or cap the total amount of functional elements that could possibly exist in the genome since elements that do things at the molecular level can be missed by comparative genomics.

Furthermore, much of the apparent junk DNA is involved in epigenetic regulation and appears to be necessary for the development of complex organisms.

In a 2014 paper, ENCODE researchers tried to address "the question of whether nonconserved but biochemically active regions are truly functional". They noted that in the literature, functional parts of the genome have been identified differently in previous studies depending on the approaches used. There have been three general approaches used to identify functional parts of the human genome: genetic approaches (which rely on changes in phenotype), evolutionary approaches (which rely on conservation) and biochemical approaches (which rely on biochemical testing and was used by ENCODE). All three have limitations: genetic approaches may miss functional elements that do not manifest physically on the organism, evolutionary approaches have difficulties using accurate multispecies sequence alignments since genomes of even closely related species vary considerably, and with biochemical approaches, though having high reproducibility, the biochemical signatures do not always automatically signify a function.

They noted that 70% of the transcription coverage was less than 1 transcript per cell. They noted that this "larger proportion of genome with reproducible but low biochemical signal strength and less evolutionary conservation is challenging to parse between specific functions and biological noise". Furthermore, assay resolution often is much broader than the underlying functional sites so some of the reproducibly "biochemically active but selectively neutral" sequences are unlikely to serve critical functions, especially those with lower-level biochemical signal. To this they added, "However, we also acknowledge substantial limitations in our current detection of constraint, given that some human-specific functions are essential but not conserved and that disease-relevant regions need not be selectively constrained to be functional." On the other hand, they argued that the 12–15% fraction of human DNA under functional constraint, as estimated by a variety of extrapolative evolutionary methods, may still be an underestimate. They concluded that in contrast to evolutionary and genetic evidence, biochemical data offer clues about both the molecular function served by underlying DNA elements and the cell types in which they act. Ultimately genetic, evolutionary, and biochemical approaches can all be used in a complementary way to identify regions that may be functional in human biology and disease.

Some critics have argued that functionality can only be assessed in reference to an appropriate null hypothesis. In this case, the null hypothesis would be that these parts of the genome are non-functional and have properties, be it on the basis of conservation or biochemical activity, that would be expected of such regions based on our general understanding of molecular evolution and biochemistry. According to these critics, until a region in question has been shown to have additional features, beyond what is expected of the null hypothesis, it should provisionally be labelled as non-functional.

Functions of Noncoding DNA

Many noncoding DNA sequences have important biological functions as indicated by comparative genomics studies that report some regions of noncoding DNA that are highly conserved, sometimes on time-scales representing hundreds of millions of years, implying that these noncoding regions are under strong evolutionary pressure and positive selection. For example, in the genomes of humans and mice, which diverged from a common ancestor 65–75 million years ago, protein-coding DNA sequences account for only about 20% of conserved DNA, with the remaining

80% of conserved DNA represented in noncoding regions. Linkage mapping often identifies chromosomal regions associated with a disease with no evidence of functional coding variants of genes within the region, suggesting that disease-causing genetic variants lie in the noncoding DNA. The significance of noncoding DNA mutations in cancer was explored in April 2013.

Noncoding genetic polymorphisms have also been shown to play a role in infectious disease susceptibility, such as hepatitis C. Moreover, noncoding genetic polymorphisms were shown to contribute to susceptibility to Ewing sarcoma - a highly aggressive pediatric bone cancer.

Some specific sequences of noncoding DNA may be features essential to chromosome structure, centromere function and homolog recognition in meiosis.

According to a comparative study of over 300 prokaryotic and over 30 eukaryotic genomes, eukaryotes appear to require a minimum amount of non-coding DNA. This minimum amount can be predicted using a growth model for regulatory genetic networks, implying that it is required for regulatory purposes. In humans the predicted minimum is about 5% of the total genome.

There is evidence that a significant proportion (over 10%) of 32 mammalian genomes may function through the formation of specific RNA secondary structures. The study used comparative genomics to identify compensatory DNA mutations that maintain RNA base-pairings, a distinctive feature of RNA molecules. Over 80% of the genomic regions presenting evolutionary evidence of RNA structure conservation do not present strong DNA sequence conservation.

Protection of the Genome

Noncoding DNA separate genes from each other with long gaps, so mutation in one gene or part of a chromosome, for example deletion or insertion, does not have the frame shift effect (as in "frameshift mutation") on the whole chromosome. When genome complexity is relatively high, like in the case of human genome, not only between different genes, but also inside many genes, there are gaps of introns to protect the entire coding segment and minimise the changes caused by mutation.

It has been suggested that non-coding DNA may serve to decrease the chance of gene disruption during chromosomal crossover.

Genetic Switches

Some noncoding DNA sequences are genetic "switches" that regulate when and where genes are expressed. For example, a long non-coding RNA (lncRNA) molecule was recently found to assist in preventing breast cancer, by stopping a genetic switch from getting stuck.

Regulation of Gene Expression

Some noncoding DNA sequences determine the expression levels of various genes. In addition, some noncoding variants have been associated to variation in expression levels of mRNAs through mapping of Expression quantitative trait loci (eQTLs). When eQTLs are combined with Genome-wide association study (GWAS), some independent risk variants for the same disease impacted on expression of the same mRNAs, or on mRNAs with similar functions.

Transcription Factor Sites

Some noncoding DNA sequences determine where transcription factors attach. A transcription factor is a protein that binds to specific non-coding DNA sequences, thereby controlling the flow (or transcription) of genetic information from DNA to mRNA. Transcription factors act at very different locations on the genomes of different people.

Operators

An operator is a segment of DNA to which a repressor binds. A repressor is a DNA-binding protein that regulates the expression of one or more genes by binding to the operator and blocking the attachment of RNA polymerase to the promoter, thus preventing transcription of the genes. This blocking of expression is called repression.

Enhancers

An enhancer is a short region of DNA that can be bound with proteins (trans-acting factors), much like a set of transcription factors, to enhance transcription levels of genes in a gene cluster.

Silencers

A silencer is a region of DNA that inactivates gene expression when bound by a regulatory protein. It functions in a very similar way as enhancers, only differing in the inactivation of genes.

Promoters

A promoter is a region of DNA that facilitates transcription of a particular gene. Promoters are typically located near the genes they regulate.

Insulators

A genetic insulator is a boundary element that plays two distinct roles in gene expression, either as an enhancer-blocking code, or rarely as a barrier against condensed chromatin. An insulator in a DNA sequence is comparable to a linguistic word divider such as a comma (,) in a sentence, because the insulator indicates where an enhanced or repressed sequence ends.

Uses of Noncoding DNA

Noncoding DNA and Evolution

Shared sequences of apparently non-functional DNA are a major line of evidence of common descent.

Pseudogene sequences appear to accumulate mutations more rapidly than coding sequences due to a loss of selective pressure. This allows for the creation of mutant alleles that incorporate new functions that may be favored by natural selection; thus, pseudogenes can serve as raw material for evolution and can be considered "protogenes".

Long Range Correlations

A statistical distinction between coding and noncoding DNA sequences has been found. It has been observed that nucleotides in non-coding DNA sequences display long range power law correlations while coding sequences do not.

Forensic Anthropology

Police sometimes gather DNA as evidence for purposes of forensic identification. As described in *Maryland v. King*, a 2013 U.S. Supreme Court decision:

"The current standard for forensic DNA testing relies on an analysis of the chromosomes located within the nucleus of all human cells. "The DNA material in chromosomes is composed of 'coding' and 'noncoding' regions. The coding regions are known as genes and contain the information necessary for a cell to make proteins. . . . Non-protein coding regions . . . are not related directly to making proteins, [and] have been referred to as 'junk' DNA." The adjective "junk" may mislead the lay person, for in fact this is the DNA region used with near certainty to identify a person.

RNA

Ribonucleic acid (RNA) is a polymeric molecule essential in various biological roles in coding, decoding, regulation, and expression of genes. RNA and DNA are nucleic acids, and, along with proteins and carbohydrates, constitute the four major macromolecules essential for all known forms of life. Like DNA, RNA is assembled as a chain of nucleotides, but unlike DNA it is more often found in nature as a single-strand folded onto itself, rather than a paired double-strand. Cellular organisms use messenger RNA (*mRNA*) to convey genetic information (using the letters G, U, A, and C to denote the nitrogenous bases guanine, uracil, adenine, and cytosine) that directs synthesis of specific proteins. Many viruses encode their genetic information using an RNA genome.

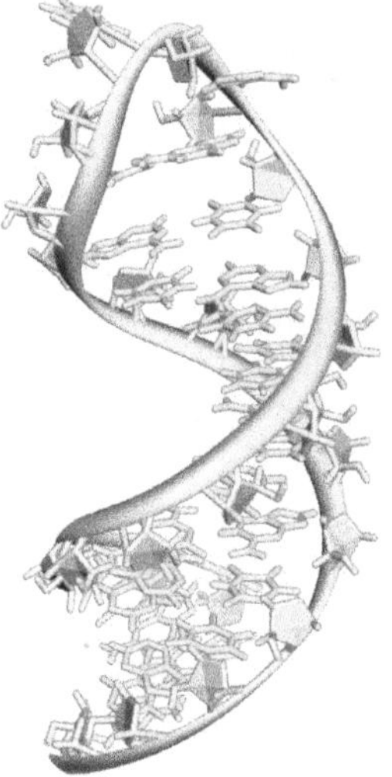

A hairpin loop from a pre-mRNA. Highlighted are the nucleobases (green) and the ribose-phosphate backbone (blue). Note that this is a single strand of RNA that folds back upon itself.

Some RNA molecules play an active role within cells by catalyzing biological reactions, controlling gene expression, or sensing and communicating responses to cellular signals. One of these active processes is protein synthesis, a universal function where RNA molecules direct the assembly of proteins on ribosomes. This process uses transfer RNA (***tRNA***) molecules to deliver amino acids to the ribosome, where ribosomal RNA (***rRNA***) then links amino acids together to form proteins.

Comparison with DNA

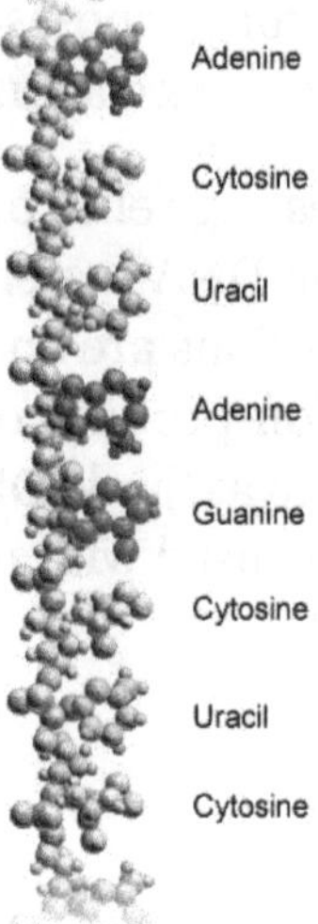

RNA Molecule

Bases in an RNA molecule.

Three-dimensional representation of the 50S ribosomal subunit. Ribosomal RNA is in ochre, proteins in blue. The active site is a small segment of rRNA, indicated in red.

The chemical structure of RNA is very similar to that of DNA, but differs in three main ways:

- Unlike double-stranded DNA, RNA is a single-stranded molecule in many of its biological roles and has a much shorter chain of nucleotides. However, RNA can, by complementary base pairing, form intrastrand (i.e., single-strand) double helixes, as in tRNA.

- While DNA contains *deoxyribose*, RNA contains *ribose* (in deoxyribose there is no hydroxyl group attached to the pentose ring in the 2' position). These hydroxyl groups make RNA less stable than DNA because it is more prone to hydrolysis.
- The complementary base to adenine in DNA is thymine, whereas in RNA, it is uracil, which is an unmethylated form of thymine.

Like DNA, most biologically active RNAs, including mRNA, tRNA, rRNA, snRNAs, and other non-coding RNAs, contain self-complementary sequences that allow parts of the RNA to fold and pair with itself to form double helices. Analysis of these RNAs has revealed that they are highly structured. Unlike DNA, their structures do not consist of long double helices, but rather collections of short helices packed together into structures akin to proteins. In this fashion, RNAs can achieve chemical catalysis (like enzymes). For instance, determination of the structure of the ribosome—an enzyme that catalyzes peptide bond formation—revealed that its active site is composed entirely of RNA.

Structure

Each nucleotide in RNA contains a ribose sugar, with carbons numbered 1' through 5'. A base is attached to the 1' position, in general, adenine (A), cytosine (C), guanine (G), or uracil (U). Adenine and guanine are purines, cytosine and uracil are pyrimidines. A phosphate group is attached to the 3' position of one ribose and the 5' position of the next. The phosphate groups have a negative charge each, making RNA a charged molecule (polyanion). The bases form hydrogen bonds between cytosine and guanine, between adenine and uracil and between guanine and uracil. However, other interactions are possible, such as a group of adenine bases binding to each other in a bulge, or the GNRA tetraloop that has a guanine–adenine base-pair.

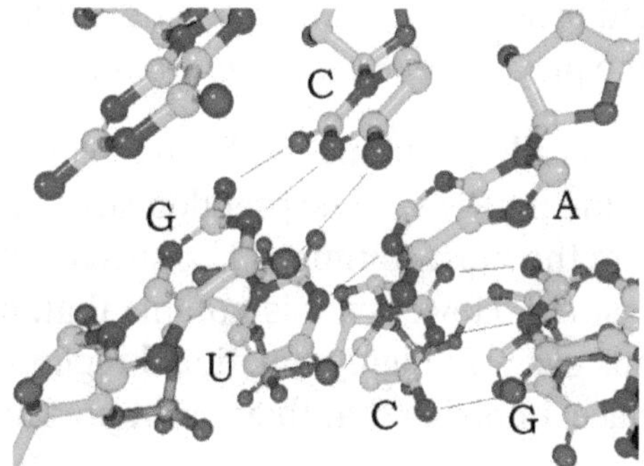

Watson-Crick base pairs in a siRNA (hydrogen atoms are not shown)

base
(here: guanine)
ribose
5'→3'
direction
phosphate

Chemical structure of RNA

An important structural feature of RNA that distinguishes it from DNA is the presence of a hydroxyl group at the 2' position of the ribose sugar. The presence of this functional group causes the helix to mostly adopt the A-form geometry, although in single strand dinucleotide contexts, RNA can rarely also adopt the B-form most commonly observed in DNA. The A-form geometry results in a very deep and narrow major groove and a shallow and wide minor groove. A second consequence of the presence of the 2'-hydroxyl group is that in conformationally flexible regions of an RNA molecule (that is, not involved in formation of a double helix), it can chemically attack the adjacent phosphodiester bond to cleave the backbone.

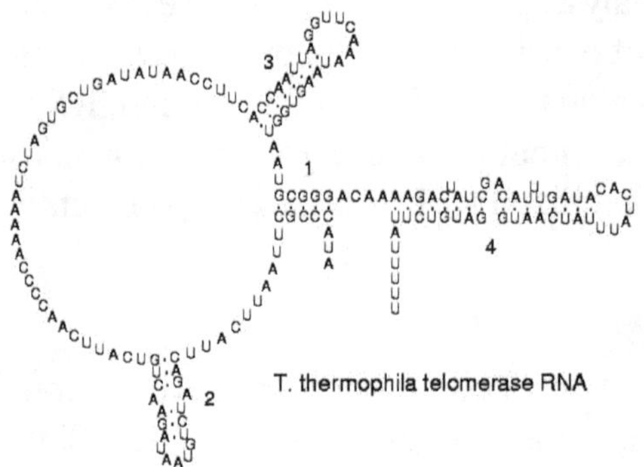

Secondary structure of a telomerase RNA.

RNA is transcribed with only four bases (adenine, cytosine, guanine and uracil), but these bases and attached sugars can be modified in numerous ways as the RNAs mature. Pseudouridine (Ψ), in which the linkage between uracil and ribose is changed from a C–N bond to a C–C bond, and ribothymidine (T) are found in various places (the most notable ones being in the TΨC loop of tRNA). Another notable modified base is hypoxanthine, a deaminated adenine base whose nucleoside is called inosine (I). Inosine plays a key role in the wobble hypothesis of the genetic code.

There are more than 100 other naturally occurring modified nucleosides, The greatest structural diversity of modifications can be found in tRNA, while pseudouridine and nucleosides with 2'-O-methylribose often present in rRNA are the most common. The specific roles of many of these modifications in RNA are not fully understood. However, it is notable that, in ribosomal RNA, many of the post-transcriptional modifications occur in highly functional regions, such as the peptidyl transferase center and the subunit interface, implying that they are important for normal function.

The functional form of single-stranded RNA molecules, just like proteins, frequently requires a specific tertiary structure. The scaffold for this structure is provided by secondary structural elements that are hydrogen bonds within the molecule. This leads to several recognizable "domains" of secondary structure like hairpin loops, bulges, and internal loops. Since RNA is charged, metal ions such as Mg^{2+} are needed to stabilise many secondary and tertiary structures.

The naturally occurring enantiomer of RNA is D-RNA composed of D-ribonucleotides. All chirality centers are located in the D-ribose. By the use of L-ribose or rather L-ribonucleotides, L-RNA can be synthesized. L-RNA is much more stable against degradation by RNase.

Like other structured biopolymers such as proteins, one can define topology of a folded RNA molecule. This is often done based on arrangement of intra-chain contacts within a folded RNA, termed as circuit topology.

Synthesis

Synthesis of RNA is usually catalyzed by an enzyme—RNA polymerase—using DNA as a template, a process known as transcription. Initiation of transcription begins with the binding of the enzyme to a promoter sequence in the DNA (usually found "upstream" of a gene). The DNA double helix is unwound by the helicase activity of the enzyme. The enzyme then progresses along the template strand in the 3' to 5' direction, synthesizing a complementary RNA molecule with elongation occurring in the 5' to 3' direction. The DNA sequence also dictates where termination of RNA synthesis will occur.

Primary transcript RNAs are often modified by enzymes after transcription. For example, a poly(A) tail and a 5' cap are added to eukaryotic pre-mRNA and introns are removed by the spliceosome.

There are also a number of RNA-dependent RNA polymerases that use RNA as their template for synthesis of a new strand of RNA. For instance, a number of RNA viruses (such as poliovirus) use this type of enzyme to replicate their genetic material. Also, RNA-dependent RNA polymerase is part of the RNA interference pathway in many organisms.

Types of RNA

Overview

Messenger RNA (mRNA) is the RNA that carries information from DNA to the ribosome, the sites of protein synthesis (translation) in the cell. The coding sequence of the mRNA determines the amino acid sequence in the protein that is produced. However, many RNAs do not code for protein (about 97% of the transcriptional output is non-protein-coding in eukaryotes).

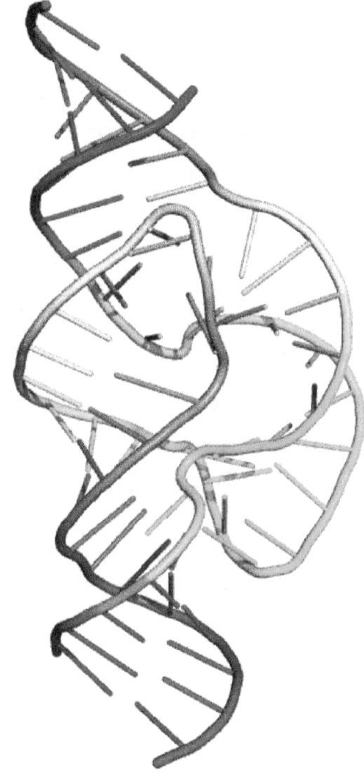

Structure of a hammerhead ribozyme, a ribozyme that cuts RNA

These so-called non-coding RNAs ("ncRNA") can be encoded by their own genes (RNA genes), but can also derive from mRNA introns. The most prominent examples of non-coding RNAs are transfer RNA (tRNA) and ribosomal RNA (rRNA), both of which are involved in the process of translation. There are also non-coding RNAs involved in gene regulation, RNA processing and other roles. Certain RNAs are able to catalyse chemical reactions such as cutting and ligating other RNA molecules, and the catalysis of peptide bond formation in the ribosome; these are known as ribozymes.

In Length

According to the length of RNA chain, RNA includes small RNA and long RNA. Usually, small RNAs are <200 nt in length, and long RNA are >200 nt. Long RNAs, also called large RNAs, mainly include long non-coding RNA (lncRNA) and mRNA. Small RNAs mainly include 5.8S ribosomal RNA (rRNA), 5S rRNA, transfer RNA (tRNA), microRNA (miRNA), small interfering RNA (siRNA), small nucleolar RNA (snoRNAs), Piwi-interacting RNA (piRNA), tRNA-derived small RNA (tsRNA) and small rDNA-derived RNA (srRNA).

In Translation

Messenger RNA (mRNA) carries information about a protein sequence to the ribosomes, the protein synthesis factories in the cell. It is coded so that every three nucleotides (a codon) corresponds to one amino acid. In eukaryotic cells, once precursor mRNA (pre-mRNA) has been transcribed from DNA, it is processed to mature mRNA. This removes its introns—non-coding sections of the pre-mRNA. The mRNA is then exported from the nucleus to the cytoplasm, where it is bound to ribosomes and translated into its corresponding protein form with the help of tRNA. In prokaryotic cells, which do not have nucleus and cytoplasm compartments, mRNA can bind to ribosomes while it is being transcribed from DNA. After a certain amount of time the message degrades into its component nucleotides with the assistance of ribonucleases.

Transfer RNA (tRNA) is a small RNA chain of about 80 nucleotides that transfers a specific amino acid to a growing polypeptide chain at the ribosomal site of protein synthesis during translation. It has sites for amino acid attachment and an anticodon region for codon recognition that binds to a specific sequence on the messenger RNA chain through hydrogen bonding.

Ribosomal RNA (rRNA) is the catalytic component of the ribosomes. Eukaryotic ribosomes contain four different rRNA molecules: 18S, 5.8S, 28S and 5S rRNA. Three of the rRNA molecules are synthesized in the nucleolus, and one is synthesized elsewhere. In the cytoplasm, ribosomal RNA and protein combine to form a nucleoprotein called a ribosome. The ribosome binds mRNA and carries out protein synthesis. Several ribosomes may be attached to a single mRNA at any time. Nearly all the RNA found in a typical eukaryotic cell is rRNA.

Transfer-messenger RNA (tmRNA) is found in many bacteria and plastids. It tags proteins encoded by mRNAs that lack stop codons for degradation and prevents the ribosome from stalling.

Regulatory RNAs

Several types of RNA can downregulate gene expression by being complementary to a part of an mRNA or a gene's DNA. MicroRNAs (miRNA; 21-22 nt) are found in eukaryotes and act through

RNA interference (RNAi), where an effector complex of miRNA and enzymes can cleave complementary mRNA, block the mRNA from being translated, or accelerate its degradation.

While small interfering RNAs (siRNA; 20-25 nt) are often produced by breakdown of viral RNA, there are also endogenous sources of siRNAs. siRNAs act through RNA interference in a fashion similar to miRNAs. Some miRNAs and siRNAs can cause genes they target to be methylated, thereby decreasing or increasing transcription of those genes. Animals have Piwi-interacting RNAs (piRNA; 29-30 nt) that are active in germline cells and are thought to be a defense against transposons and play a role in gametogenesis.

Many prokaryotes have CRISPR RNAs, a regulatory system similar to RNA interference. Antisense RNAs are widespread; most downregulate a gene, but a few are activators of transcription. One way antisense RNA can act is by binding to an mRNA, forming double-stranded RNA that is enzymatically degraded. There are many long noncoding RNAs that regulate genes in eukaryotes, one such RNA is Xist, which coats one X chromosome in female mammals and inactivates it.

An mRNA may contain regulatory elements itself, such as riboswitches, in the 5' untranslated region or 3' untranslated region; these cis-regulatory elements regulate the activity of that mRNA. The untranslated regions can also contain elements that regulate other genes.

In RNA Processing

Many RNAs are involved in modifying other RNAs. Introns are spliced out of pre-mRNA by spliceosomes, which contain several small nuclear RNAs (snRNA), or the introns can be ribozymes that are spliced by themselves. RNA can also be altered by having its nucleotides modified to nucleotides other than A, C, G and U. In eukaryotes, modifications of RNA nucleotides are in general directed by small nucleolar RNAs (snoRNA; 60-300 nt), found in the nucleolus and cajal bodies. snoRNAs associate with enzymes and guide them to a spot on an RNA by basepairing to that RNA. These enzymes then perform the nucleotide modification. rRNAs and tRNAs are extensively modified, but snRNAs and mRNAs can also be the target of base modification. RNA can also be methylated.

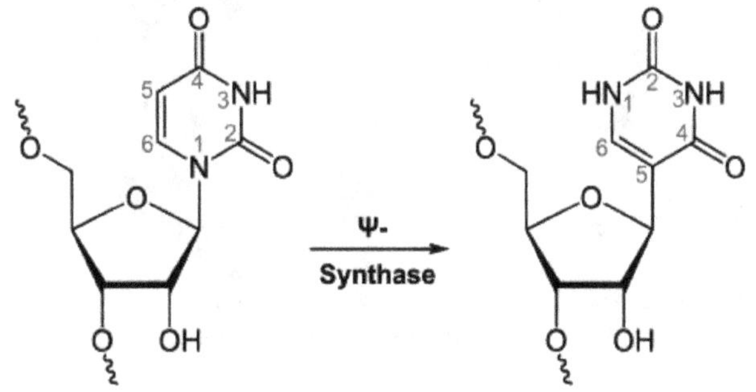

Uridine to pseudouridine is a common RNA modification.

RNA Genomes

Like DNA, RNA can carry genetic information. RNA viruses have genomes composed of RNA that encodes a number of proteins. The viral genome is replicated by some of those proteins, while other proteins protect the genome as the virus particle moves to a new host cell. Viroids are another group of pathogens, but they consist only of RNA, do not encode any protein and are replicated by a host plant cell's polymerase.

In Reverse Transcription

Reverse transcribing viruses replicate their genomes by reverse transcribing DNA copies from their RNA; these DNA copies are then transcribed to new RNA. Retrotransposons also spread by copying DNA and RNA from one another, and telomerase contains an RNA that is used as template for building the ends of eukaryotic chromosomes.

Double-stranded RNA

Double-stranded RNA (dsRNA) is RNA with two complementary strands, similar to the DNA found in all cells. dsRNA forms the genetic material of some viruses (double-stranded RNA viruses). Double-stranded RNA such as viral RNA or siRNA can trigger RNA interference in eukaryotes, as well as interferon response in vertebrates.

Circular RNA

Recently, it was shown that there is a single stranded covalently closed, *i.e.* circular form of RNA expressed throughout the animal and plant kingdom. circRNAs are thought to arise via a "back-splice" reaction where the spliceosome joins a downstream donor to an upstream acceptor splice site. So far the function of circRNAs is largely unknown, although for few examples a microRNA sponging activity has been demonstrated.

Key Discoveries in RNA Biology

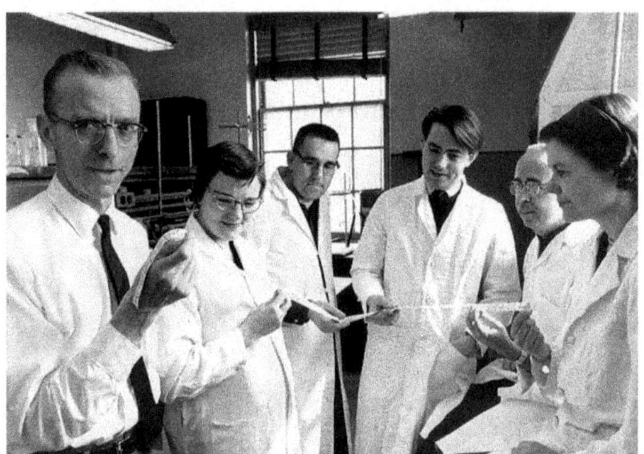

Robert W. Holley, left, poses with his research team.

Research on RNA has led to many important biological discoveries and numerous Nobel Prizes. Nucleic acids were discovered in 1868 by Friedrich Miescher, who called the material 'nuclein' since it was found in the nucleus. It was later discovered that prokaryotic cells, which do not have a nucleus, also contain nucleic acids. The role of RNA in protein synthesis was suspected already in 1939. Severo Ochoa won the 1959 Nobel Prize in Medicine (shared with Arthur Kornberg) after he discovered an enzyme that can synthesize RNA in the laboratory. However, the enzyme discovered by Ochoa (polynucleotide phosphorylase) was later shown to be responsible for RNA degradation, not RNA synthesis. In 1956 Alex Rich and David Davies hybridized two separate strands of RNA to form the first crystal of RNA whose structure could be determined by X-ray crystallography.

The sequence of the 77 nucleotides of a yeast tRNA was found by Robert W. Holley in 1965, winning Holley the 1968 Nobel Prize in Medicine (shared with Har Gobind Khorana and Marshall Nirenberg). In 1967, Carl Woese hypothesized that RNA might be catalytic and suggested that the earliest forms of life (self-replicating molecules) could have relied on RNA both to carry genetic information and to catalyze biochemical reactions—an RNA world.

During the early 1970s, retroviruses and reverse transcriptase were discovered, showing for the first time that enzymes could copy RNA into DNA (the opposite of the usual route for transmission of genetic information). For this work, David Baltimore, Renato Dulbecco and Howard Temin were awarded a Nobel Prize in 1975. In 1976, Walter Fiers and his team determined the first complete nucleotide sequence of an RNA virus genome, that of bacteriophage MS2.

In 1977, introns and RNA splicing were discovered in both mammalian viruses and in cellular genes, resulting in a 1993 Nobel to Philip Sharp and Richard Roberts. Catalytic RNA molecules (ribozymes) were discovered in the early 1980s, leading to a 1989 Nobel award to Thomas Cech and Sidney Altman. In 1990, it was found in *Petunia* that introduced genes can silence similar genes of the plant's own, now known to be a result of RNA interference.

At about the same time, 22 nt long RNAs, now called microRNAs, were found to have a role in the development of *C. elegans*. Studies on RNA interference gleaned a Nobel Prize for Andrew Fire and Craig Mello in 2006, and another Nobel was awarded for studies on the transcription of RNA to Roger Kornberg in the same year. The discovery of gene regulatory RNAs has led to attempts to develop drugs made of RNA, such as siRNA, to silence genes.

Evolution

In March 2015, complex DNA and RNA organic compounds of life, including uracil, cytosine and thymine, were reportedly formed in the laboratory under outer space conditions, using starter chemicals, such as pyrimidine, found in meteorites. Pyrimidine, like polycyclic aromatic hydrocarbons (PAHs), the most carbon-rich chemical found in the Universe, may have been formed in red giants or in interstellar dust and gas clouds, according to the scientists.

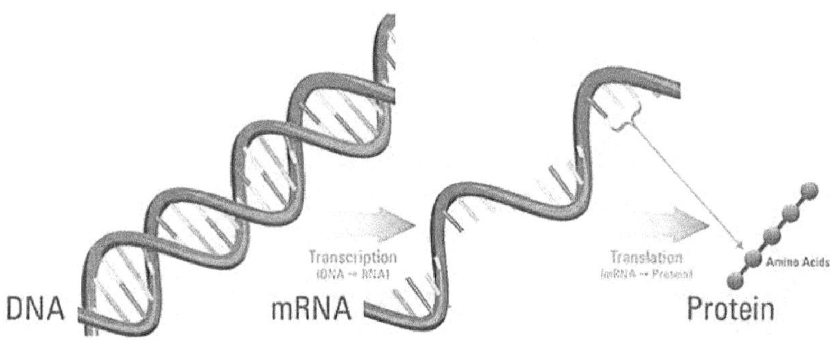

References

- Jones, Daniel (2003) [1917], Peter Roach, James Hartmann and Jane Setter, eds., English Pronouncing Dictionary, Cambridge: Cambridge University Press, ISBN 3-12-539683-2 CS1 maint: Uses editors parameter (link)
- Carlson, Elof A. (2004). Mendel's Legacy: The Origin of Classical Genetics (PDF). Cold Spring Harbor, NY: Cold Spring Harbor Laboratory Press. p. 88. ISBN 978-087969675-7.
- White, M. J. D. (1973). The chromosomes (6th ed.). London: Chapman and Hall, distributed by Halsted Press, New York. p. 28. ISBN 0-412-11930-7.
- Miller, Kenneth R. (2000). "Chapter 9-3". Biology (5th ed.). Upper Saddle River, New Jersey: Prentice Hall. pp. 194–5. ISBN 0-13-436265-9.
- Alberts B, Johnson A, Lewis J, Raff M, Roberts K & Wlater P (2002). Molecular Biology of the Cell (4th ed.). Garland Science. ISBN 0-8153-3218-1. pp. 120–121.
- Walsh CP, Xu GL (2006). "Cytosine methylation and DNA repair". Curr Top Microbiol Immunol. Current Topics in Microbiology and Immunology. 301: 283–315. doi:10.1007/3-540-31390-7_11. ISBN 3-540-29114-8. PMID 16570853.
- Tani, Katsuji; Nasu, Masao (2010). "Roles of Extracellular DNA in Bacterial Ecosystems". In Kikuchi, Yo; Rykova, Elena Y. Extracellular Nucleic Acids. Springer. pp. 25–38. ISBN 978-3-642-12616-1.
- Ito T (2003). "Nucleosome assembly and remodelling". Curr Top Microbiol Immunol. Current Topics in Microbiology and Immunology. 274: 1–22. doi:10.1007/978-3-642-55747-7_1. ISBN 978-3-540-44208-0. PMID 12596902.
- Houdebine LM (2007). "Transgenic animal models in biomedical research". Methods Mol Biol. 360: 163–202. doi:10.1385/1-59745-165-7:163. ISBN 1-59745-165-7. PMID 17172731.
- Baldi, Pierre; Brunak, Soren (2001). Bioinformatics: The Machine Learning Approach. MIT Press. ISBN 978-0-262-02506-5. OCLC 45951728.
- Gusfield, Dan. Algorithms on Strings, Trees, and Sequences: Computer Science and Computational Biology. Cambridge University Press, 15 January 1997. ISBN 978-0-521-58519-4.
- Mount DM (2004). Bioinformatics: Sequence and Genome Analysis (2 ed.). Cold Spring Harbor, NY: Cold Spring Harbor Laboratory Press. ISBN 0-87969-712-1. OCLC 55106399.
- Costa, Fabrico (2012). "7 Non-coding RNAs, Epigenomics, and Complexity in Human Cells". In Morris, Kevin V. Non-coding RNAs and Epigenetic Regulation of Gene Expression: Drivers of Natural Selection. Caister Academic Press. ISBN 1904455948.
- Carey, Nessa (2015). Junk DNA: A Journey Through the Dark Matter of the Genome. Columbia University Press. ISBN 9780231170840.
- Morris, Kevin, ed. (2012). Non-Coding RNAs and Epigenetic Regulation of Gene Expression: Drivers of Natural Selection. Norfolk, UK: Caister Academic Press. ISBN 1904455948.
- Berg JM; Tymoczko JL; Stryer L (2002). Biochemistry (5th ed.). WH Freeman and Company. pp. 118–19, 781–808. ISBN 0-7167-4684-0. OCLC 179705944.
- Barciszewski J; Frederic B; Clark C (1999). RNA biochemistry and biotechnology. Springer. pp. 73–87. ISBN 0-7923-5862-7. OCLC 52403776.
- Jankowski JAZ; Polak JM (1996). Clinical gene analysis and manipulation: Tools, techniques and troubleshooting. Cambridge University Press. p. 14. ISBN 0-521-47896-0. OCLC 33838261.
- Söll D; RajBhandary U (1995). TRNA: Structure, biosynthesis, and function. ASM Press. p. 165. ISBN 1-55581-073-X. OCLC 183036381.

- Cooper GC; Hausman RE (2004). The Cell: A Molecular Approach (3rd ed.). Sinauer. pp. 261–76, 297, 339–44. ISBN 0-87893-214-3. OCLC 174924833.
- Wirta W (2006). Mining the transcriptome – methods and applications. Stockholm: School of Biotechnology, Royal Institute of Technology. ISBN 91-7178-436-5. OCLC 185406288.
- Wagner EG; Altuvia S; Romby P (2002). "Antisense RNAs in bacteria and their genetic elements". Adv Genet. Advances in Genetics. 46: 361–98. doi:10.1016/S0065-2660(02)46013-0. ISBN 9780120176465. PMID 11931231.

3

Breeding: An Overview

Breeding is the function of reproduction and is concerned with particularly animals and plants. Some of the aspects of breeding discussed are captive breeding, seasonal breeder, breeding pair, inbreeding, outcrossing, backcrossing and culling. This section is an overview of the subject matter incorporating all the major aspects of breeding.

Captive Breeding

USFWS staff with two red wolf pups bred in captivity

Captive breeding is the process of breeding animals in controlled environments within well-defined settings, such as wildlife reserves, zoos and other commercial and noncommercial conservation facilities. Sometimes the process includes the release of individual organisms to the wild, when there is sufficient natural habitat to support new individuals or when the threat to the species in the wild is lessened. Captive breeding programs facilitate biodiversity and may save species from extinction. Release programs have the potential for diluting genetic diversity and fitness.

History

Captive breeding has been successful in the past. The Pere David's deer was successfully saved through captive breeding programs after almost being hunted to extinction in China.

Captive-breeding is employed by modern conservationists, and has saved a wide variety of species from extinction, ranging from birds (e.g., the pink pigeon), mammals (e.g., the pygmy hog), reptiles (e.g., the Round Island boa) and amphibians (e.g., poison arrow frogs). Their efforts were successful in reintroducing the Arabian oryx (under the auspices of the Fauna and Flora Preservation Society), in 1963. The Przewalski's horse was also successfully reintroduced in the wild after being bred in captivity.

Coordination

The breeding of endangered species is coordinated by cooperative breeding programs containing international studbooks and coordinators, who evaluate the roles of individual animals and institutions from a global or regional perspective. These studbooks contain information on birth date, gender, location, and lineage (if known), which helps determine survival and reproduction rates, number of founders of the population, and inbreeding coefficients. A species coordinator reviews the information in studbooks and determines a breeding strategy that would produce most advantageous offspring.

If two compatible animals are found at different zoos, the animals may be transported for mating, but this is stressful, which could in turn make mating less likely. However, this is still a popular breeding method among European zoological organizations. Artificial fertilization (by shipping semen) is another option, but male animals can experience stress during semen collection, and the same goes for females during the artificial insemination procedure. Furthermore, this approach yields lower-quality semen because shipping requires the life of the sperm to be extended for the transit time.

There are regional programmes for the conservation of endangered species:

- Americas: Species Survival Plan SSP (Association of Zoos and Aquariums AZA, Canadian Association of Zoos and Aquariums CAZA)
- Europe: European Endangered Species Programme EEP (European Association of Zoos and Aquaria EAZA)
- Australasia: Australasian Species Management Program ASMP (Zoo and Aquarium Association ZAA)
- Africa: African Preservation Program APP (African Association of Zoological Gardens and Aquaria PAAZAB)
- Japan: Conservation activities of Japanese Association of Zoos and Aquariums JAZA
- South Asia: Conservation activities of South Asian Zoo Association for Regional Cooperation SAZARC
- South East Asia: Conservation activities of South East Asian Zoos Association SEAZA

Challenges

Captive breeding techniques are usually difficult to implement for highly mobile species such as migratory birds like cranes and fishes like Hilsa.

Conservation biologists define endangered species as one that is likely to become extinct in the near future and is designated as endangered on the IUCN Red List. Conservation seeks to eliminate threats from human activities such as habitat loss and fragmentation, hunting, fishing, pollution, predation, disease, and parasitism.

Genetics

Recall that endangered species are those on the verge of extinction and so are more than a very small population. A risk of captive breeding includes inbreeding, i.e., mating between two closely related individuals as a result of a small gene pool. Inbreeding may lead to decreased disease immunity and phenotypic abnormalities. With the possibility of inbreeding, populations may undergo genetic drift, where genes have the potential to disappear completely, not only reducing genetic variation but also undermining natural selection by pressuring the remaining population and their predators. In the case of captive breeding prior to reintroduction into the wild, modelling works indicate that the duration of programs (i.e., time from the foundation of the captive population to the last release event) is an important determinant of reintroduction success. Success is maximized for intermediate project duration allowing to release a sufficient number of individuals, while maintaining the number of generations of relaxed selection in captivity to an acceptable level. Over a sufficient number of generations, however, inbred populations can regain "normal" genetic diversity.

For example, since the 1970s the Matschie's tree-kangaroo, an endangered species, has been bred in captivity. The Tree Kangaroo Species Survival Plan (TKSSP) was established in 1992 to help with the management of Association of Zoos and Aquariums (AZA). TKSSP's annual breeding recommendations to preserve genetic diversity are based on mean kinship strategy (in order to retain adaptive potential and avoid inbreeding's disadvantages). In order to evaluate a captive breeding program's performance (in maintaining genetic diversity), researchers compare the genetic diversity of the captive breeding population to the wild population. According to McGreezy et al. (2010), "AZA Matschie tree kangaroo's haplotype diversity was almost two times lower than wild Matschie tree kangaroos". This difference with allele frequencies shows the changes that can happen over time, like genetic drift and mutation, when a species is taken out of its natural habitat.

Another example is the cheetah, the least genetically variable felid species. This makes it very difficult to pair animals in a way that would increase genetic diversity because all cheetahs are essentially genetically identical. It is also possible that human practices are causing even more of an inbreeding depression in the handling of the cats in zoos than exists within the wild populations. Note that although the cheetah has undergone genetic drift in the form of bottlenecks thousands of years ago, they seem to experience few of inbreeding's detrimental effects.

Behaviour Changes

Captive breeding can contribute to behavioural problems in animals that are subsequently released because they are unable to hunt or forage for food leading to starvation, possibly because the young animals spent the critical learning period in captivity. Released animals often do not avoid predators and are not able to find ample shelter for themselves and may die. Golden lion tamarin mothers often die in the wild before having offspring because they cannot climb and forage. This leads to continuing population declines despite reintroduction as the species are unable to produce viable offspring. Training can improve anti-predator skills, but its effectiveness varies.

Loss of Habitat

Another challenge with captive breeding is the habitat loss that occurs while they are in captivity being bred (though it is occurring even before they are captured). This may make release of the species nonviable if there is no habitat left to support larger populations.

Climate change and invasive species are threatening an increasing number of species with extinction. A decrease in population size can reduce genetic diversity, which detracts from a population's ability to adapt in a changing environment. In this way extinction risk is related to loss of genetic polymorphism, which is a difference in DNA sequence among individuals, groups or populations. Conservation programs can now obtain measurements of genetic diversity at functionally important genes thanks to advances in technology.

Assortative Mating

A study on mice has found that after captive breeding had been in place for multiple generations and these mice were "released" to breed with wild mice, that the captive-born mice bred amongst themselves instead of with the wild mice. This suggests that captive breeding may affect mating preferences, and has implications for the success of a reintroduction program.

Successes

The De Wildt Cheetah and Wildlife Centre, established in South Africa in 1971, has a cheetah captive breeding program. Between 1975 and 2005, 242 litters were born with a total of 785 cubs. The survival rate of cubs was 71.3% for the first twelve months and 66.2% for older cubs, validating the fact that cheetahs can be bred successfully (and their endangerment decreased). It also indicated that failure in other breeding habitats may be due to "poor" sperm morphology.

Wild Tasmanian devils have declined by 90% due to a transmissible cancer called Devil Facial Tumor Disease. A captive insurance population program has started, but the captive breeding rates at the moment are lower than they need to be. Keeley, Fanson, Masters, and McGreevy (2012) sought to "increase our understanding of the estrous cycle of the devil and elucidate potential causes of failed male-female pairings" by examining temporal patterns of fecal progestogen and corticosterone metabolite concentrations. They found that the majority of unsuccessful females were captive-born, suggesting that if the species' survival depended solely on captive breeding, the population would probably disappear.

In 2010, the Oregon Zoo found that Columbia Basin pygmy rabbit pairings based on familiarity and preferences resulted in a significant increase in breeding success.

New Technologies

The major histocompatibility complex (MHC) is a genome region that is emerging as an exciting research field. Researchers found that genes that code for MHC affect the ability of certain species, such as *Batrachochytrium dendrobatidis*, to resist certain infections because the MHC has a mediating effect on the interaction between the body's immune cells with other body cells. Measuring polymorphism at these genes can serve as an indirect measure of a population's immunological fitness. Captive breeding programs that selectively breed for disease-resistant genes may facilitate successful reintroductions.

There have also been recent advances in captive breeding programs with the use of induced pluripotent stem cell (iPSC) technology, which has been tested on endangered species. Scientists hope that they can convert stem cells into germ cells in order to diversify the gene pools of threatened species. Healthy mice have been born with this technology. iPSC may one day be used to treat captive animals with diseases.

Seasonal Breeder

Seasonal breeders are animal species that successfully mate only during certain times of the year. These times of year allow for the births at a time optimal for the survival of the young in terms of factors such as ambient temperature, food and water availability, and even changes in the predation behaviors of other species. Related sexual interest and behaviors are expressed and accepted only during this period. Female seasonal breeders will have one or more estrus cycles only when she is "in season" or fertile and receptive to mating. At other times of the year, they will be anestrus. Unlike reproductive cyclicity, seasonality is described in both males and females. Male seasonal breeders may exhibit changes in testosterone levels, testes weight, and fertility depending on the time of year.

Seasonal breeders are distinct from opportunistic breeders, which mate whenever the conditions of their environment become favorable, and continuous breeders like humans that mate year-round.

Breeding Season

The breeding season is the most suitable season, usually with favourable conditions and abundant food and water, for breeding. Abiotic factors such as the timing of seasonal rains and winds can also play an important role in breeding onset and success.

Many species breed in colonies or large communities which is known as communal breeding. It is common to see large congregations of these species in particular favourable locations in their breeding seasons. These breeding colonies and their location are generally protected by wildlife conservation laws to keep the species from going extinct. Some species have evolved for communal breeding in large breeding colonies and can not breed in smaller numbers or pairs alone. These species can be threatened by imminent extinction if they are hunted on their breeding grounds or if their breeding colonies are destroyed. The Passenger pigeon is a famous example of probably the most numerous land bird on the American continent which had evolved for communal breeding that became extinct due to large scale hunting in its communal breeding grounds during the breeding season and its inability to breed in smaller numbers.

Physiology

The hypothalamus is considered to be the central control for reproduction. Hence, factors that determine when a seasonal breeder will be ready for mating affect this tissue. This is achieved specifically through changes in the production of the hormone GnRH. GnRH in turn transits to the pituitary where it promotes the secretion of the gonadotropins LH and FSH, both pituitary hormones critical for reproductive function and behavior, into the bloodstream. Changes in gonadotropin secretion initiate the end of anestrus in females.

Factors that Determine Time of Fertility

Photoperiod

When a seasonal breeder is ready for mating is strongly regulated by length of day (photoperiod) and thus season. Photoperiod likely affects the seasonal breeder through changes in melatonin secretion by the pineal gland that ultimately alter GnRH release by the hypothalamus.

Hence, seasonal breeders can be divided into groups based on when they are fertile. "Long day" breeders cycle when days get longer (spring) and are in anestrus in fall and winter. Some animals that are long day breeders includes; Ring-tailed lemurs, Horses, Hamsters, Groundhogs, and Minks. "Short day" breeders cycle when the length of daylight shortens (fall) and are in anestrus in spring and summer. The decreased light during the fall decreases the firing of the retinal nerves, in turn decreasing the excitation of the superior cervical ganglion, which then decreases the inhibition of the pineal gland, finally resulting in an increase in melatonin. This increase in melatonin results in an increase in GnRH and subsequently an increase in the hormones LH and FSH, which stimulate cyclicity. Some animals that are short day breeders includes; Sheeps, Goats, Foxes, Red Deers, Elks, Mooses, and Mice.

Day length variations with latitude can also impact breeding. For instance, sheep and goats in tropical climes may breed throughout the year while those in more polar arctic areas may have a shortened season. In addition, females are generally more sensitive to changes in day length. For instance, unlike mares, stallions remain fertile year-round, suffering only some declines in sexual behavior and sperm production out of season.

Other Factors

Other factors that affect breeding time include the presence of a ready and available mate. For instance, the presence of a fertile buck will induce an estrus cycle in a doe shortly after introduction. Further environmental factors can include nutrition, chemosensory and hormonal cues. Weight and age are other factors.

Breeding Pair

Breeding pair is a pair of animals which cooperate over time to produce offspring with some form of a bond between the individuals. For example, many birds mate for a breeding season or sometimes for life. They may share some or all of the tasks involved: building a nest, incubating the eggs and feeding and protecting the young. The term is not generally used when a male has a harem (zoology) of females, such as with mountain gorillas.

True breeding pairs are usually found only in vertebrates, but there are notable exceptions, such as the Lord Howe Island stick insect. True breeding pairs are rare in amphibians or reptiles, but fairly common with fish (e.g. discus) and especially birds. Breeding pair arrangements are rare in mammals, where the prevailing patterns are either that the male and female only meet for copulation (e.g. brown bear) or that dominant males have a harem (zoology) of females (e.g. walrus).

Selective Breeding

A Belgian Blue cow. The defect in the breed's myostatin gene is maintained through linebreeding and is responsible for its accelerated lean muscle growth.

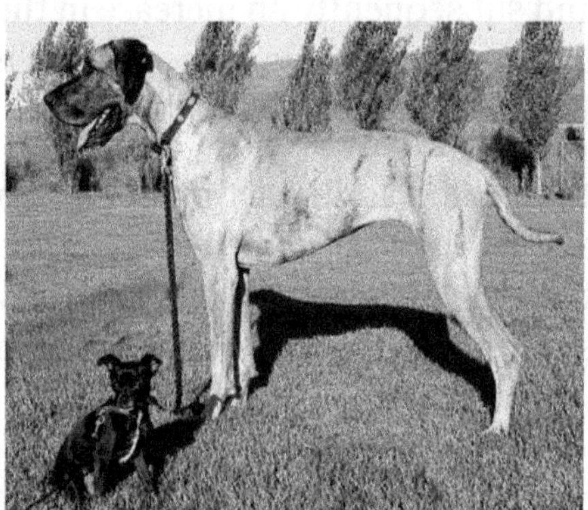

This Chihuahua mix and Great Dane shows the wide range of dog breed sizes created using selective breeding.

Selective breeding transformed teosinte's few fruitcases (left) into modern maize's rows of exposed kernels (right).

Selective breeding (also called artificial selection) is the process by which humans use animal breeding and plant breeding to selectively develop particular phenotypic traits (characteristics) by choosing which typically animal or plant males and females will sexually reproduce and have offspring together. Domesticated animals are known as breeds, normally bred by a professional breeder, while domesticated plants are known as varieties, cultigens, or cultivars. Two purebred animals of different breeds produce a crossbreed, and crossbred plants are called hybrids. Flowers, vegetables and fruit-trees may be bred by amateurs and commercial or non-commercial professionals: major crops are usually the provenance of the professionals.

There are two approaches or types of artificial selection, or selective breeding. First is the traditional "breeder's approach" in which the breeder or experimenter applies "a known amount of selection to a single phenotypic trait" by examining the chosen trait and choosing to breed only those that exhibit higher or "extreme values" of that trait. The second is called "controlled natural selection," which is essentially natural selection in a controlled environment. In this, the breeder does not choose which individuals being tested "survive or reproduce," as he or she could in the traditional approach. There are also "selection experiments," which is a third approach and these are conducted in order to determine the "strength of natural selection in the wild." However, this is more often an observational approach as opposed to an experimental approach.

In animal breeding, techniques such as inbreeding, linebreeding, and outcrossing are utilized. In plant breeding, similar methods are used. Charles Darwin discussed how selective breeding had been successful in producing change over time in his 1859 book, *On the Origin of Species*. Its first chapter discusses selective breeding and domestication of such animals as pigeons, cats, cattle, and dogs. Darwin used artificial selection as a springboard to introduce and support the theory of natural selection.

The deliberate exploitation of selective breeding to produce desired results has become very common in agriculture and experimental biology.

Selective breeding can be unintentional, e.g., resulting from the process of human cultivation; and it may also produce unintended – desirable or undesirable – results. For example, in some grains, an increase in seed size may have resulted from certain ploughing practices rather than from the intentional selection of larger seeds. Most likely, there has been an interdependence between natural and artificial factors that have resulted in plant domestication.

History

Selective breeding of both plants and animals has been practiced since early prehistory; key species such as wheat, rice, and dogs have been significantly different from their wild ancestors for millennia, and maize, which required especially large changes from teosinte, its wild form, was selectively bred in Mesoamerica. Selective breeding was practiced by the Romans. Treatises as much as 2,000 years old give advice on selecting animals for different purposes, and these ancient works cite still older authorities, such as Mago the Carthaginian. The notion of selective breeding was later expressed by the Persian Muslim polymath Abu Rayhan Biruni in the 11th century. He noted the idea in his book titled *India*, which included various examples.

The agriculturist selects his corn, letting grow as much as he requires, and tearing out the remainder. The forester leaves those branches which he perceives to be excellent, whilst he cuts away all others. The bees kill those of their kind who only eat, but do not work in their beehive.

—Abu Rayhan Biruni, India

Selective breeding was established as a scientific practice by Robert Bakewell during the British Agricultural Revolution in the 18th century. Arguably, his most important breeding program was with sheep. Using native stock, he was able to quickly select for large, yet fine-boned sheep, with long, lustrous wool. The Lincoln Longwool was improved by Bakewell, and in turn the Lincoln was used to develop the subsequent breed, named the New (or Dishley) Leicester. It was hornless and had a square, meaty body with straight top lines.

These sheep were exported widely, including to Australia and North America, and have contributed to numerous modern breeds, despite the fact that they fell quickly out of favor as market preferences in meat and textiles changed. Bloodlines of these original New Leicesters survive today as the English Leicester (or Leicester Longwool), which is primarily kept for wool production.

Bakewell was also the first to breed cattle to be used primarily for beef. Previously, cattle were first and foremost kept for pulling ploughs as oxen, but he crossed long-horned heifers and a Westmoreland bull to eventually create the Dishley Longhorn. As more and more farmers followed his lead, farm animals increased dramatically in size and quality. In 1700, the average weight of a bull sold for slaughter was 370 pounds (168 kg). By 1786, that weight had more than doubled to 840 pounds (381 kg). However, after his death, the Dishley Longhorn was replaced with short-horn versions.

He also bred the Improved Black Cart horse, which later became the Shire horse.

Charles Darwin coined the term 'selective breeding'; he was interested in the process as an illustration of his proposed wider process of natural selection. Darwin noted that many domesticated animals and plants had special properties that were developed by intentional animal and plant breeding from individuals that showed desirable characteristics, and discouraging the breeding of individuals with less desirable characteristics.

Darwin used the term "artificial selection" twice in the 1859 first edition of his work *On the Origin of Species*, in Chapter IV: Natural Selection, and in Chapter VI: Difficulties on Theory:

Slow though the process of selection may be, if feeble man can do much by his powers of artificial selection, I can see no limit to the amount of change, to the beauty and infinite complexity of the co-adaptations between all organic beings, one with another and with their physical conditions of life, which may be effected in the long course of time by nature's power of selection.

—Charles Darwin, On the Origin of Species

We are profoundly ignorant of the causes producing slight and unimportant variations; and we are immediately made conscious of this by reflecting on the differences in the breeds of our domesticated animals in different countries,—more especially in the less civilized countries where there has been but little artificial selection.

Animal Breeding

Animals with homogeneous appearance, behavior, and other characteristics are known as particular breeds, and they are bred through culling animals with particular traits and selecting for further breeding those with other traits. Purebred animals have a single, recognizable breed, and purebreds with recorded lineage are called pedigreed. Crossbreeds are a mix of two purebreds, whereas mixed breeds are a mix of several breeds, often unknown. Animal breeding begins with breeding stock, a group of animals used for the purpose of planned breeding. When individuals are looking to breed animals, they look for certain valuable traits in purebred stock for a certain purpose, or may intend to use some type of crossbreeding to produce a new type of stock with different, and, it is presumed, superior abilities in a given area of endeavor. For example, to breed chickens, a breeder typically intends to receive eggs, meat, and new, young birds for further reproduction. Thus, the breeder has to study different breeds and types of chickens and analyze what can be expected from a certain set of characteristics before he or she starts breeding them. Therefore, when purchasing initial breeding stock, the breeder seeks a group of birds that will most closely fit the purpose intended.

Three generations of "Westies" in a village in Fife, Scotland

Purebred breeding aims to establish and maintain stable traits, that animals will pass to the next generation. By "breeding the best to the best," employing a certain degree of inbreeding, considerable culling, and selection for "superior" qualities, one could develop a bloodline superior in certain respects to the original base stock. Such animals can be recorded with a breed registry, the organization that maintains pedigrees and/or stud books. However, single-trait breeding, breeding for only one trait over all others, can be problematic. In one case mentioned by animal behaviorist Temple Grandin, roosters bred for fast growth or heavy muscles did not know how to perform typical rooster courtship dances, which alienated the roosters from hens and led the roosters to kill the hens after reproducing with them.

The observable phenomenon of hybrid vigor stands in contrast to the notion of breed purity. However, on the other hand, indiscriminate breeding of crossbred or hybrid animals may also result in degradation of quality. Studies in evolutionary physiology, behavioral genetics, and other areas of organismal biology have also made use of deliberate selective breeding, though longer generation times and greater difficulty in breeding can make such projects challenging in vertebrates.

Plant Breeding

Plant breeding has been used for thousands of years, and began with the domestication of wild plants into uniform and predictable agricultural cultigens. High-yielding varieties have been particularly important in agriculture.

Researchers at the USDA have selectively bred carrots with a variety of colors.

Selective plant breeding is also used in research to produce transgenic animals that breed "true" (i.e., are homozygous) for artificially inserted or deleted genes.

Selective Breeding in Aquaculture

Selective breeding in aquaculture holds high potential for the genetic improvement of fish and shellfish. Unlike terrestrial livestock, the potential benefits of selective breeding in aquaculture were not realized until recently. This is because high mortality led to the selection of only a few broodstock, causing inbreeding depression, which then forced the use of wild broodstock. This was evident in selective breeding programs for growth rate, which resulted in slow growth and high mortality.

Control of the reproduction cycle was one of the main reasons as it is a requisite for selective breeding programmes. Artificial reproduction was not achieved because of the difficulties in hatching or feeding some farmed species such as eel and yellowtail farming. A suspected reason associated with the late realisation of success in selective breeding programs in aquaculture was the education of the concerned people – researchers, advisory personnel and fish farmers. The education of fish biologists paid less attention to quantitative genetics and breeding plans.

Another was the failure of documentation of the genetic gains in successive generations. This in turn led to failure in quantifying economic benefits that successful selective breeding programs produce. Documentation of the genetic changes was considered important as they help in fine tuning further selection schemes.

Quality Traits in Aquaculture

Aquaculture species are reared for particular traits such as growth rate, survival rate, meat quality, resistance to diseases, age at sexual maturation, fecundity, shell traits like shell size, shell colour, etc.

- Growth rate – growth rate is normally measured as either body weight or body length. This trait is of great economic importance for all aquaculture species as faster growth rate speeds up the turnover of production. Improved growth rates show that farmed animals utilize their feed more efficiently through a correlated response.
- Survival rate – survival rate may take into account the degrees of resistance to diseases. This may also see the stress response as fish under stress are highly vulnerable to diseases. The stress fish experience could be of biological, chemical or environmental influence.
- Meat quality – the quality of fish is of great economic importance in the market. Fish quality usually takes into account size, meatiness, and percentage of fat, colour of flesh, taste, shape of the body, ideal oil and omega-3 content.
- Age at sexual maturation – The age of maturity in aquaculture species is another very important attribute for farmers as during early maturation the species divert all their energy to gonad production affecting growth and meat production and are more susceptible to health problems (Gjerde 1986).
- Fecundity – As the fecundity in fish and shellfish is usually high it is not considered as a major trait for improvement. However, selective breeding practices may consider the size of the egg and correlate it with survival and early growth rate.

Finfish Response to Selection

Salmonids

Gjedrem (1979) showed that selection of Atlantic salmon (*Salmo salar*) led to an increase in body weight by 30% per generation. A comparative study on the performance of select Atlantic salmon with wild fish was conducted by AKVAFORSK Genetics Centre in Norway. The traits, for which the selection was done included growth rate, feed consumption, protein retention, energy retention, and feed conversion efficiency. Selected fish had a twice better growth rate, a 40% higher feed intake, and an increased protein and energy retention. This led to an overall 20% better Fed Conversion Efficiency as compared to the wild stock. Atlantic salmon have also been selected for resistance to bacterial and viral diseases. Selection was done to check resistance to Infectious Pancreatic Necrosis Virus (IPNV). The results showed 66.6% mortality for low-resistant species whereas the high-resistant species showed 29.3% mortality compared to wild species.

Rainbow trout (*S. gairdneri*) was reported to show large improvements in growth rate after 7–10 generations of selection. Kincaid et al. (1977) showed that growth gains by 30% could be achieved by selectively breeding rainbow trout for three generations. A 7% increase in growth was recorded per generation for rainbow trout by Kause et al. (2005).

In Japan, high resistance to IPNV in rainbow trout has been achieved by selectively breeding the stock. Resistant strains were found to have an average mortality of 4.3% whereas 96.1% mortality was observed in a highly sensitive strain.

Coho salmon (*Oncorhynchus kisutch*) increase in weight was found to be more than 60% after four generations of selective breeding. In Chile, Neira et al. (2006) conducted experiments on early spawning dates in coho salmon. After selectively breeding the fish for four generations, spawning dates were 13–15 days earlier.

Cyprinids

Selective breeding programs for the Common carp (*Cyprinus carpio*) include improvement in growth, shape and resistance to disease. Experiments carried out in the USSR used crossings of broodstocks to increase genetic diversity and then selected the species for traits like growth rate, exterior traits and viability, and/or adaptation to environmental conditions like variations in temperature. Kirpichnikov *et al.* (1974) and Babouchkine (1987) selected carp for fast growth and tolerance to cold, the Ropsha carp. The results showed a 30–40% to 77.4% improvement of cold tolerance but did not provide any data for growth rate. An increase in growth rate was observed in the second generation in Vietnam. Moav and Wohlfarth (1976) showed positive results when selecting for slower growth for three generations compared to selecting for faster growth. Schaperclaus (1962) showed resistance to the dropsy disease wherein selected lines suffered low mortality (11.5%) compared to unselected (57%).

Channel Catfish

Growth was seen to increase by 12–20% in selectively bred *Iictalurus punctatus*. More recently, the response of the Channel Catfish to selection for improved growth rate was found to be approximately 80%, i.e., an average of 13% per generation.

Shellfish Response to Selection

Oysters

Selection for live weight of Pacific oysters showed improvements ranging from 0.4% to 25.6% compared to the wild stock. Sydney-rock oysters (*Saccostrea commercialis*) showed a 4% increase after one generation and a 15% increase after two generations. Chilean oysters (*Ostrea chilensis*), selected for improvement in live weight and shell length showed a 10–13% gain in one generation. Bonamia ostrea is a protistan parasite that causes catastrophic losses (nearly 98%) in European flat oyster *Ostrea edulis* L. This protistan parasite is endemic to three oyster-regions in Europe. Selective breeding programs show that *O. edulis* susceptibility to the infection differs across oyster strains in Europe. A study carried out by Culloty et al. showed that 'Rossmore' oysters in Cork harbour, Ireland had better resistance compared to other Irish strains. A selective breeding program at Cork harbour uses broodstock from 3- to 4-year-old survivors and is further controlled until a viable percentage reaches market size. Over the years 'Rossmore' oysters have shown to develop lower prevalence to *B. ostreae* infection and percentage mortality. Ragone Calvo et al. (2003) selectively bred the eastern oyster, *Crassostrea virginica*, for resistance against co-occurring parasites *Haplosporidium nelson* (MSX) and *Perkinsus marinus* (Dermo). They achieved dual resistance to the disease in four generations of selective breeding. The oysters showed higher growth and survival rates and low susceptibility to the infections. At the end of the experiment, artificially selected *C. virginica* showed a 34–48% higher survival rate.

Penaeid Shrimps

Selection for growth in Penaeid shrimps yielded successful results. A selective breeding program for *Litopenaeus stylirostris* saw an 18% increase in growth after the fourth generation and 21% growth after the fifth generation. *Marsupenaeus japonicas* showed a 10.7% increase in growth after the first generation. Argue et al. (2002) conducted a selective breeding program on the Pacific White Shrimp, *Litopenaeus vannamei* at The Oceanic Institute, Waimanalo, USA from 1995 to 1998. They reported significant responses to selection compared to the unselected control shrimps. After one generation, a 21% increase was observed in growth and 18.4% increase in survival to TSV. The Taura Syndrome Virus (TSV) causes mortalities of 70% or more in shrimps. C.I. Oceanos S.A. in Colombia selected the survivors of the disease from infected ponds and used them as parents for the next generation. They achieved satisfying results in two or three generations wherein survival rates approached levels before the outbreak of the disease. The resulting heavy losses (up to 90%) caused by Infectious hypodermal and haematopoietic necrosis virus (IHHNV) caused a number of shrimp farming industries started to selectively breed shrimps resistant to this disease. Successful outcomes led to development of Super Shrimp, a selected line of *L. stylirostris* that is resistant to IHHNV infection. Tang et al. (2000) confirmed this by showing no mortalities in IHHNV- challenged Super Shrimp post larvae and juveniles.

Aquatic Species Versus Terrestrial Livestock

Selective breeding programs for aquatic species provide better outcomes compared to terrestrial livestock. This higher response to selection of aquatic farmed species can be attributed to the following:

- High fecundity in both sexes fish and shellfish enabling higher selection intensity.
- Large phenotypic and genetic variation in the selected traits.

Selective breeding in aquaculture provide remarkable economic benefits to the industry, the primary one being that it reduces production costs due to faster turnover rates. This is because of faster growth rates, decreased maintenance rates, increased energy and protein retention, and better feed efficiency. Applying such genetic improvement program to aquaculture species will increase productivity to meet the increasing demands of growing populations.

Advantages and Disadvantages

Selective breeding is a direct way to determine if a specific trait can "evolve in response to selection." A single-generation method of breeding is not as accurate or direct. The process is also more practical and easier to understand than sibling analysis. The former tests "differences between line means" while the latter is dependent upon "variance and covariance components." Essentially, selective breeding is better for traits such as physiology and behavior that are hard to measure because it requires fewer individuals to test than single-generation testing.

However, there are disadvantages to this process. Because a single experiment done in selective breeding cannot be used to assess an entire group of "genetic variances and covariances," individual experiments must be done for every individual trait. Also, because of the necessity of selective breeding experiments to require maintaining the organisms tested in a lab or greenhouse, it is impractical to use this breeding method on many organisms. Controlled mating instances are difficult to carry out in this case and this is a necessary component of selective breeding.

Inbreeding

Inbreeding is the production of offspring from the mating or breeding of individuals or organisms that are closely related genetically. By analogy, the term is used in human reproduction, but more commonly refers to the genetic disorders and other consequences that may arise from incestuous sexual relationships and consanguinity.

Inbreeding results in homozygosity, which can increase the chances of offspring being affected by recessive or deleterious traits. This generally leads to a decreased biological fitness of a population (called inbreeding depression), which is its ability to survive and reproduce. An individual who inherits such deleterious traits is referred to as *inbred*. The avoidance of expression of such deleterious recessive alleles caused by inbreeding, via inbreeding avoidance mechanisms, is the main selective reason for outcrossing. Crossbreeding between populations also often has positive effects on fitness-related traits.

Inbreeding is a technique used in selective breeding. In livestock breeding, breeders may use inbreeding when, for example, trying to establish a new and desirable trait in the stock, but will need to watch for undesirable characteristics in offspring, which can then be eliminated through further selective breeding or culling. Inbreeding is used to reveal deleterious recessive alleles, which can then be eliminated through assortative breeding or through culling. In plant breeding, inbred lines are used as stocks for the creation of hybrid lines to make use of the effects of heterosis. Inbreeding in plants also occurs naturally in the form of self-pollination.

Overview

Offspring of biologically related persons are subject to the possible effect of inbreeding, such as congenital birth defects. The chances of such disorders are increased when the biological parents are more closely related. This is because such pairings have a 25% probability of producing homozygous zygotes, resulting in offspring with two recessive alleles, which can produce disorders when these alleles are deleterious. Because most recessive alleles are rare in populations, it is unlikely that two unrelated marriage partners will both be carriers of the same deleterious allele. However, because close relatives share a large fraction of their alleles, the probability that any such deleterious allele is inherited from the common ancestor through both parents is increased dramatically. It should also be noted that, in terms of probability, for each homozygous recessive individual formed, there is an equal chance of producing a homozygous dominant individual, one completely devoid of the harmful allele. Contrary to common belief, inbreeding does not in itself alter allele frequencies, but rather increases the relative proportion of homozygotes to heterozygotes. However, because the increased proportion of deleterious homozygotes exposes the allele to natural selection, in the long run its frequency decreases more rapidly in inbred populations. In the short term, incestuous reproduction is expected to increase the number of spontaneous abortions of zygotes, perinatal deaths, and postnatal offspring with birth defects. The advantages of inbreeding may be the result of a tendency to preserve the structures of alleles interacting at different loci that have been adapted together by a common selective history.

Malformations or harmful traits can stay within a population due to a high homozygosity rate, and this will cause a population to become fixed for certain traits, like having too many bones in an area, like the vertebral column of wolves on Isle Royale or having cranial abnormalities, such as in Northern elephant seals, where their cranial bone length in the lower mandibular tooth row has changed. Having a high homozygosity rate is problematic for a population because it will unmask recessive deleterious alleles generated by mutations, reduce heterozygote advantage, and it is detrimental to the survival of small, endangered animal populations. When deleterious recessive alleles are unmasked due to the increased homozygosity generated by inbreeding, this can cause inbreeding depression. The authors think that it is possible that the severity of inbreeding depression can be diminished if natural selection can purge such alleles from populations during inbreeding. If inbreeding depression can be diminished by natural selection then some traits, harmful or not, can be reduced and change the future outlook on small, endangered populations.

There may also be other deleterious effects besides those caused by recessive diseases. Thus, similar immune systems may be more vulnerable to infectious diseases.

Inbreeding history of the population should also be considered when discussing the variation in the severity of inbreeding depression between and within species. With persistent inbreeding, there is evidence that shows that inbreeding depression becoming less severe. This is associated with the unmasking and elimination of severely deleterious recessive alleles. However, inbreeding depression is not a temporary phenomenon because this elimination of deleterious recessive alleles will never be complete. Eliminating slightly deleterious mutations through inbreeding under moderate selection is not as effective. Fixation of alleles most likely occurs through Muller's ratchet, when an asexual population's genome accumulates deleterious mutations that are irreversible.

Despite all its disadvantages, inbreeding can also have a variety of advantages, such as reducing the cost of sex, reducing the recombination load, and allowing the expression of recessive advantageous phenotypes. It has been proposed that under circumstances when the advantages of inbreeding outweigh the disadvantages, preferential breeding within small groups could be promoted, potentially leading to speciation.

Genetic Disorders

Autosomal recessive disorders occur in individuals who have two copies of an allele for a particular recessive genetic mutation. Except in certain rare circumstances, such as new mutations or uniparental disomy, both parents of an individual with such a disorder will be carriers of the gene. These carriers do not display any signs of the mutation and may be unaware that they carry the mutated gene. Since relatives share a higher proportion of their genes than do unrelated people, it is more likely that related parents will both be carriers of the same recessive allele, and therefore their children are at a higher risk of inheriting an autosomal recessive genetic disorder. The extent to which the risk increases depends on the degree of genetic relationship between the parents; the risk is greater when the parents are close relatives and lower for relationships between more distant relatives, such as second cousins, though still greater than for the general population.

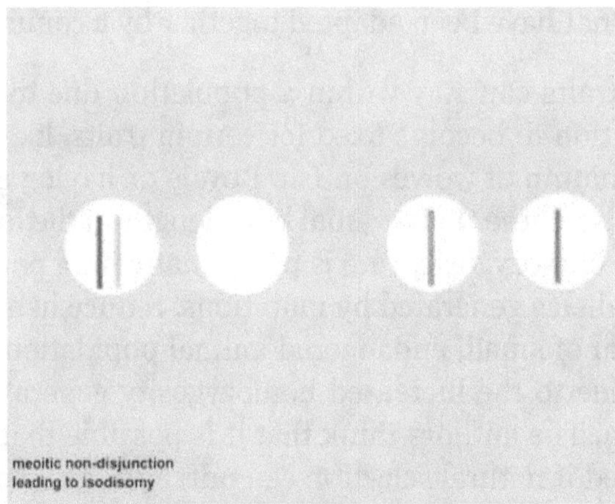

Animation of uniparental isodisomy

Children of parent-child or sibling-sibling unions are at an increased risk compared to cousin-cousin unions. Inbreeding may result in a greater than expected phenotypic expression of deleterious recessive alleles within a population. As a result, first-generation inbred individuals are more likely to show physical and health defects, including:

- Reduced fertility both in litter size and sperm viability
- Increased genetic disorders
- Fluctuating facial asymmetry
- Lower birth rate
- Higher infant mortality and child mortality
- Smaller adult size
- Loss of immune system function
- Increased cardiovascular risks

The isolation of a small population for a period of time can lead to inbreeding within that population, resulting in increased genetic relatedness between breeding individuals. Inbreeding depression can also occur in a large population if individuals tend to mate with their relatives, instead of mating randomly.

Many individuals in the first generation of inbreeding will never live to reproduce. Over time, with isolation, such as a population bottleneck caused by purposeful (assortative) breeding or natural environmental factors, the deleterious inherited traits are culled.

Island species are often very inbred, as their isolation from the larger group on a mainland allows natural selection to work on their population. This type of isolation may result in the formation of race or even speciation, as the inbreeding first removes many deleterious genes, and permits the expression of genes that allow a population to adapt to an ecosystem. As the adaptation becomes more pronounced, the new species or race radiates from its entrance into the new space, or dies out if it cannot adapt and, most importantly, reproduce.

The reduced genetic diversity, for example due to a bottleneck will unavoidably increase inbreeding for the entire population. This may mean that a species may not be able to adapt to changes in environmental conditions. Each individual will have similar immune systems, as immune systems are genetically based. When a species becomes endangered, the population may fall below a minimum whereby the forced interbreeding between the remaining animals will result in extinction.

Natural breedings include inbreeding by necessity, and most animals only migrate when necessary. In many cases, the closest available mate is a mother, sister, grandmother, father, brother, or grandfather. In all cases, the environment presents stresses to remove from the population those individuals who cannot survive because of illness.

There was an assumption that wild populations do not inbreed; this is not what is observed in some cases in the wild. However, in species such as horses, animals in wild or feral conditions often drive off the young of both sexes, thought to be a mechanism by which the species instinctively avoids some of the genetic consequences of inbreeding. In general, many mammal species, including humanity's closest primate relatives, avoid close inbreeding possibly due to the deleterious effects.

Examples

Although there are several examples of inbred populations of wild animals, the negative consequences of this inbreeding are poorly documented. In the South American sea lion, there was concern that recent population crashes would reduce genetic diversity. Historical analysis indicated that a population expansion from just two matrilineal lines was responsible for most of the individuals within the population. Even so, the diversity within the lines allowed great variation in the gene pool that may help to protect the South American sea lion from extinction.

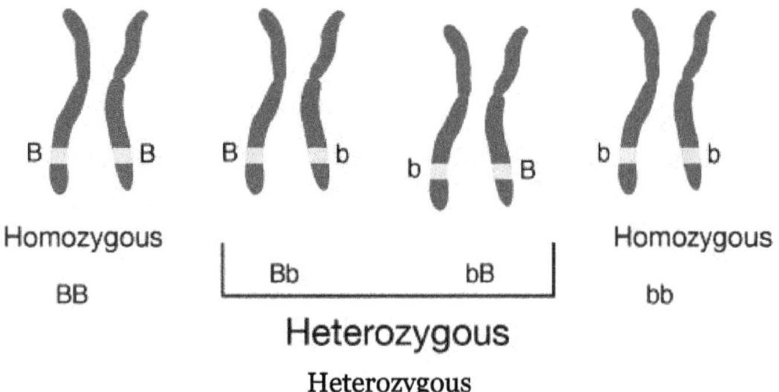

Heterozygous

In lions, prides are often followed by related males in bachelor groups. When the dominant male is killed or driven off by one of these bachelors, a father may be replaced by his son. There is no mechanism for preventing inbreeding or to ensure outcrossing. In the prides, most lionesses are related to one another. If there is more than one dominant male, the group of alpha males are usually related. Two lines are then being "line bred". Also, in some populations, such as the Crater lions, it is known that a population bottleneck has occurred. Researchers found far greater genetic heterozygosity than expected. In fact, predators are known for low genetic variance, along with most of the top portion of the trophic levels of an ecosystem. Additionally, the alpha males of two neighboring prides can be from the same litter; one brother may come to acquire leadership over another's pride, and subsequently mate with his 'nieces' or cousins. However, killing another male's cubs, upon the takeover, allows the new selected gene complement of the incoming alpha male to prevail over the previous male. There are genetic assays being scheduled for lions to determine their genetic diversity. The preliminary studies show results inconsistent with the outcrossing paradigm based on individual environments of the studied groups.

In Central California, Sea Otters were thought to have been driven to extinction due to over hunting, until a colony of about 30 breeding pairs was discovered in the Big Sur region in the 1930s. Since then, the population has grown and spread along the central Californian coast to around 2,000 individuals, a level that has remained stable for over a decade. Population growth is limited by the fact that all Californian Sea Otters are descended from the isolated colony, resulting in inbreeding.

Cheetahs are another example of inbreeding. Thousands of years ago the cheetah went through a population bottleneck that reduced its population dramatically so the animals that are alive today are all related to one another. A consequence from inbreeding for this species has been high juvenile mortality, low fecundity, and poor breeding success.

In a study on an island population of song sparrows, individuals that were inbred showed significantly lower survival rates than outbred individuals during a severe winter weather related population crash. These studies show that inbreeding depression and ecological factors have an influence on survival.

Measures of Inbreeding

A measure of inbreeding of an individual A is the probability *F*(A) that both alleles in one locus are derived from the same allele in an ancestor. These two identical alleles that are both derived from a common ancestor are said to be identical by descent. This probability F(A) is called the "coefficient of inbreeding".

Another useful measure that describes the extent to which two individuals are related (say individuals A and B) is their coancestry coefficient f(A,B), which gives the probability that one randomly selected allele from A and another randomly selected allele from B are identical by descent. This is also denoted as the kinship coefficient between A and B.

A particular case is the self-coancestry of individual A with itself, f(A,A), which is the probability that taking one random allele from A and then, independently and with replacement, another random allele also from A, both are identical by descent. Since they can be identical by descent by sampling the same allele or by sampling both alleles that happen to be identical by descent, we have $f(A,A) = 1/2 + F(A)/2$.

Both the inbreeding and the coancestry coefficients can be defined for specific individuals or as average population values. They can be computed from genealogies or estimated from the population size and its breeding properties, but all methods assume no selection and are limited to neutral alleles.

There are several methods to compute this percentage. The two main ways are the path method and the tabular method.

Typical coancestries between relatives are as follows:

- Father/daughter, mother/son or brother/sister → 25% ($^1/_4$)
- Grandfather/granddaughter or grandmother/grandson → 12.5% ($^1/_8$)
- Half-brother/half-sister, Double cousins → 12.5% ($^1/_8$)
- Uncle/niece or aunt/nephew → 12.5% ($^1/_8$)
- Great-grandfather/great-granddaughter or great-grandmother/great-grandson → 6.25% ($^1/_{16}$)
- Half-uncle/niece or half-aunt/nephew → 6.25% ($^1/_{16}$)
- First cousins → 6.25% ($^1/_{16}$)

Domestic Animals

An intensive form of inbreeding where an individual S is mated to his daughter D1, granddaughter D2 and so on, in order to maximise the percentage of S's genes in the offspring. 87.5% of D3's genes would come from S, while D4 would receive 93.75% of their genes from S.

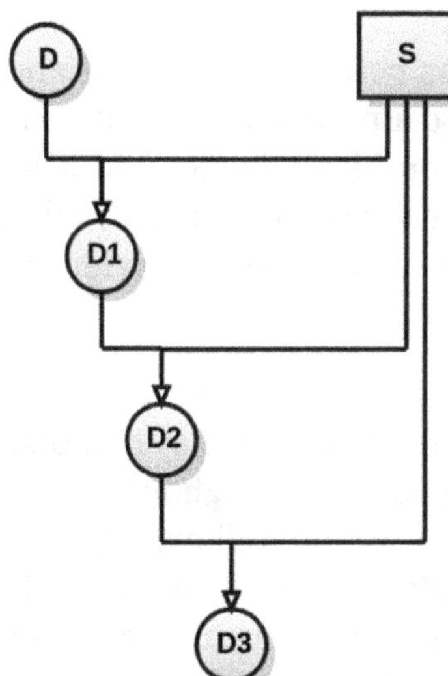

Breeding in domestic animals is primarily assortative breeding. Without the sorting of individuals by trait, a breed could not be established, nor could poor genetic material be removed. Homozygosity is the case where similar or identical alleles combine to express a trait that is not otherwise expressed (recessiveness). Inbreeding exposes recessive alleles through increasing homozygosity.

Breeders must avoid breeding from individuals that demonstrate either homozygosity or heterozygosity for disease causing alleles. The goal of preventing the transfer of deleterious alleles may be achieved by reproductive isolation, sterilization, or, in the extreme case, culling. Culling is not strictly necessary if genetics are the only issue in hand. Small animals such as cats and dogs may be sterilized, but in the case of large agricultural animals, such as cattle, culling is usually the only economic option.

The issue of casual breeders who inbreed irresponsibly is discussed in the following quotation on cattle:

Meanwhile, milk production per cow per lactation increased from 17,444 lbs to 25,013 lbs from 1978 to 1998 for the Holstein breed. Mean breeding values for milk of Holstein cows increased by 4,829 lbs during this period. High producing cows are increasingly difficult to breed and are subject to higher health costs than cows of lower genetic merit for production (Cassell, 2001).

Intensive selection for higher yield has increased relationships among animals within breed and increased the rate of casual inbreeding.

Many of the traits that affect profitability in crosses of modern dairy breeds have not been studied in designed experiments. Indeed, all crossbreeding research involving North American breeds and strains is very dated (McAllister, 2001) if it exists at all.

The BBC produced two documentaries on dog inbreeding titled Pedigree Dogs Exposed and Pedigree Dogs Exposed - Three Years On that document the negative health consequences of excessive inbreeding.

Linebreeding is a form of inbreeding. There is no clear distinction between the two terms, but linebreeding may encompass crosses between individuals and their descendants or two cousins. This method can be used to increase a particular animal's contribution to the population. While linebreeding is less likely to cause problems in the first generation than does inbreeding, over time, linebreeding can reduce the genetic diversity of a population and cause problems related to a too-small genepool that may include an increased prevalence of genetic disorders and inbreeding depression.

Outcrossing is where two unrelated individuals are crossed to produce progeny. In outcrossing, unless there is verifiable genetic information, one may find that all individuals are distantly related to an ancient progenitor. If the trait carries throughout a population, all individuals can have this trait. This is called the founder effect. In the well established breeds, that are commonly bred, a large gene pool is present. For example, in 2004, over 18,000 Persian cats were registered. A possibility exists for a complete outcross, if no barriers exist between the individuals to breed. However, it is not always the case, and a form of distant linebreeding occurs. Again it is up to the assortative breeder to know what sort of traits, both positive and negative, exist within the diversity of one breeding. This diversity of genetic expression, within even close relatives, increases the variability and diversity of viable stock.

Laboratory Animals

Systematic inbreeding and maintenance of inbred strains of laboratory mice and rats is of great importance for biomedical research. The inbreeding guarantees a consistent and uniform animal model for experimental purposes and enables genetic studies in congenic and knock-out animals. The use of inbred strains is also important for genetic studies in animal models, for example to distinguish genetic from environmental effects. The mice that are inbred typically show considerably lower survival rates.

Humans

Effects

Inbreeding increases the chances of the expression of deleterious recessive alleles by increasing homozygosity and therefore has the potential to decrease the fitness of the offspring. With continuous inbreeding, genetic variation is lost and homozygosity is increased, enabling the expression of recessive deleterious alleles in homozygotes. The inbreeding coefficient, a term used to describe the degree of inbreeding in an individual, is an estimate of the percent of homozygous alleles in the overall genome. The more biologically related the parents are, the greater the inbreeding coefficient, since their genomes have many similarities already. This overall homozygosity becomes an issue when there are deleterious recessive alleles in the gene pool of the family. By pairing chromosomes of similar genomes, the chance for these recessive alleles to pair and become homozygous greatly increases, leading to offspring with autosomal recessive disorders.

Inbreeding is especially problematic in small populations where the genetic variation is already limited. By inbreeding, individuals are further decreasing genetic variation by increasing homozygosity in the genomes of their offspring. Thus, the likelihood of deleterious recessive alleles to pair is significantly higher in a small inbreeding population than in a larger inbreeding population.

The fitness consequences of consanguineous mating have been studied since their scientific recognition by Charles Darwin in 1839. Some of the most harmful effects known from such breeding includes its effects on the mortality rate as well as on the general health of the offspring. Within the past several decades, there have been many studies to support such debilitating effects on the human organism. Specifically, inbreeding has been found to decrease fertility as a direct result of increasing homozygosity of deleterious recessive alleles. Fetuses produced by inbreeding also face a greater risk of spontaneous abortions due to inherent complications in development. Among mothers who experience stillbirths and early infant deaths, those that are inbreeding have a significantly higher chance of reaching repeated results with future offspring. Additionally, consanguineous parents possess a high risk of premature birth and producing underweight and undersized infants. Viable inbred offspring are also likely to be inflicted with physical deformities and genetically inherited diseases. Studies have confirmed an increase in several genetic disorders due to inbreeding such as blindness, hearing loss, neonatal diabetes, limb malformations, Schizophrenia and several others. Moreover, there is an increased risk for congenital heart disease depending on the inbreeding coefficient of the offspring, with significant risk accompanied by an F =.125 or higher.

Prevalence

The general negative outlook and eschewal of inbreeding that is prevalent in the Western world today holds roots from over 1500 years ago. Specifically, written documents such as the Bible illustrate that there have been laws and social customs that have called for the abstention from inbreeding. Along with cultural taboos, parental education and awareness of inbreeding consequences have played large roles in minimizing inbreeding frequencies in areas like Europe. That being so, there are less urbanized and less populated regions across the world that have shown continuity in the practice of inbreeding. This continuity is often either by choice or unavoidably due to the limitations of the geographical area. When by choice, the rate of consanguinity is highly dependent on religion and culture. Of the practicing regions, Middle Eastern and northern Africa territories show the greatest frequencies of consanguinity. The link between the high frequency and the region is primarily due to the dominance of Islamic populations, who have historically engaged in familyline relations.

- Qatar: Historically, populations of Qatar have engaged in consanguineous relationships of all kinds, leading to high risk of inheriting genetic diseases. As of 2014, around 5% of the Qatari population suffered from hereditary hearing loss; most were descendants of a consanguineous relationship.

Among these populations with high levels of inbreeding, researchers have found several disorders prevalent among inbred offspring. Specifically, in Lebanon, Saudi Arabia, Egypt, and Arabs in Israel, it has been discovered that offspring of consanguineous relationships have an increased risk of congenital malformations, congenital heart defects, congenital hydrocephalus and neural tube defects. Furthermore, among inbred children in Palestine and Lebanon, there is a positive association between consanguinity and reported cleft lip/palate cases.

Royalty and Nobility

Inter-nobility marriage was used as a method of forming political alliances among elites. These ties were often sealed only upon the birth of progeny within the arranged marriage. Thus marriage was seen as a union of lines of nobility, not as a contract between individuals as it is seen today.

Royal intermarriage was often practiced among European royal families, usually for interests of state. Over time, due to the relatively limited number of potential consorts, the gene pool of many ruling families grew progressively smaller, until all European royalty was related. This also resulted in many being descended from a certain person through many lines of descent, such as the numerous European royalty and nobility descended from the British Queen Victoria or King Christian IX of Denmark. The House of Habsburg was infamous for its inbreeding, with the Habsburg lip cited as an ill-effect, although no genetic evidence has proved the allegation. The closely related houses of Habsburg, Bourbon, Braganza and Wittelsbach also frequently engaged in first-cousin unions as well as the occasional double-cousin and uncle-niece marriages. Examples of incestuous marriages and the impact of inbreeding on royal families include:

- In ancient Egypt, royal women were believed to carry the bloodlines and so it was advantageous for a pharaoh to marry his sister or half-sister; in such cases a special combination between endogamy and polygamy is found. Normally, the old ruler's eldest son and daughter (who could be either siblings or half-siblings) became the new rulers. All rulers of the Ptolemaic dynasty uninterruptedly from Ptolemy IV (Ptolemy II married his sister but had no issue) were married to their brothers and sisters, so as to keep the Ptolemaic blood "pure" and to strengthen the line of succession. Cleopatra VII (also called Cleopatra VI) and Ptolemy XIII, who married and became co-rulers of ancient Egypt following their father's death, are the most widely known example.

- One of the most famous examples of a genetic trait aggravated by royal family intermarriage was the House of Habsburg, which inmarried particularly often and is known for the mandibular prognathism of the *Habsburger (Unter) Lippe* (otherwise known as the

'Habsburg jaw', 'Habsburg lip' or 'Austrian lip'"). This was typical for many Habsburg relatives over a period of six centuries. The condition progressed through the generations to the point that the last of the Spanish Habsburgs, Charles II of Spain, could not properly chew his food.

- Besides the jaw deformity, Charles II also had a huge number of other genetic physical, intellectual, sexual, and emotional problems. It is speculated that the simultaneous occurrence in Charles II of two different genetic disorders (combined pituitary hormone deficiency and distal renal tubular acidosis) could explain most of the complex clinical profile of this king, including his impotence/infertility, which led to the extinction of the dynasty.

Outcrossing

Outcrossing or outbreeding is the practice of introducing unrelated genetic material into a breeding line. It increases genetic diversity, thus reducing the probability of an individual being subject to disease or reducing genetic abnormalities.

Outcrossing is now the norm of most purposeful animal breeding, contrary to what is commonly believed. The outcrossing breeder intends to remove the traits by using "new blood". With dominant traits, one can still see the expression of the traits and can remove those traits whether one outcrosses, line breeds or inbreds. With recessive traits, outcrossing allows for the recessive traits to migrate across a population. The outcrossing breeder then may have individuals that have many deleterious genes that may be expressed by subsequent inbreeding. There is now a gamut of deleterious genes within each individual in many dog breeds.

Increasing the variation of genes or alleles within the gene pool may protect against extinction by stressors from the environment. For example, in this context, a recent veterinary medicine study tried to determine the genetic diversity within cat breeds.

Outcrossing is believed to be the "norm" in the wild. Outcrossing in plants is usually enforced by self-incompatibility.

Breeders inbreed within their genetic pool, attempting to maintain desirable traits and to cull those traits that are undesirable. When undesirable traits begin to appear, mates are selected to determine if a trait is recessive or dominant. Removal of the trait is accomplished by breeding two individuals known not to carry it.

Gregor Mendel used outcrossing in his experiments with flowers. He then used the resulting offspring to chart inheritance patterns, using the crossing of siblings, and backcrossing to parents to determine how inheritance functioned.

Charles Darwin, in his book "The Effects of Cross and Self-Fertilization in the Vegetable Kingdom," came to clear and definite conclusions concerning the adaptive benefit of outcrossing. For example, he stated (on page 462) that "the offspring from the union of two distinct individuals, especially if their progenitors have been subjected to very different conditions, have an immense advantage in height, weight, constitutional vigor and fertility over the self-fertilizing offspring from either one of the same parents." He thought that this observation was amply sufficient to account for outcrossing sexual reproduction. The disadvantages of self-fertilized offspring (inbreeding depression) are now thought to be largely due to the homozygous expression of deleterious recessive mutations; and the fitness advantages of outcrossed offspring are thought to be largely due to the heterozygous masking of such deleterious mutations.

Backcrossing

Backcrossing is a crossing of a hybrid with one of its parents or an individual genetically similar to its parent, in order to achieve offspring with a genetic identity which is closer to that of the parent. It is used in horticulture, animal breeding and in production of gene knockout organisms.

Backcrossed hybrids are sometimes described with acronym "BC", for example, an F1 hybrid crossed with one of its parents (or a genetically similar individual) can be termed a BC1 hybrid, and a further cross of the BC1 hybrid to the same parent (or a genetically similar individual) produces a BC2 hybrid.

Plants

Advantages

- If the recurrent parent is an elite genotype, at the end of the backcrossing programme an elite genotype is recovered
- As there is no "new" recombination, the elite combination is not lost

Disadvantages

- Works poorly for quantitative traits
- Is more restricted for recessive traits
- In practice, sections of genome from the non-recurrent parents are often still present and can have unwanted traits associated with them
- For very wide crosses, limited recombination may maintain thousands of 'alien' genes within the elite cultivar
- Many backcrosses are required to produce a new cultivar which can take many years

Natural Backcrossings

York radiate groundsel (*Senecio eboracensis*) is a naturally occurring hybrid species of Oxford ragwort (*Senecio squalidus*) and common groundsel (*Senecio vulgaris*). It is thought to have arisen from a backcrossing of the F1 hybrid with *S. vulgaris*.

Again, the pure tall (TT) and pure dwarf (tt) pea plants when crossed in the parental generation, they produce all heterozygote (Tt) tall pea plants in the first filial generation. The cross between first filial heterozygote tall (Tt) pea plant and pure tall (TT) or pure dwarf (tt) pea plant of the parental generation is also an example for the back-crossing between two plants. In this case, the filial generation formed after the back cross may have a phenotype ratio of 1:1 if the cross is made with recessive parent or else all offspring may be having phenotype of dominant trait if back cross is with parent having dominant trait. The former of these traits is also called a test cross.

Artificially Recombinant Lines

In plants, *inbred backcross lines* (IBLs) refers to lines (i.e. populations) of plants derived from the repeated backcrossing of a line with artificially recombinant DNA with the wild type, operating some kind of selection that can be phenotypical or through a molecular marker (for the production of introgression lines).

Animals

Backcrossing may be deliberately employed in animals to transfer a desirable trait in an animal of inferior genetic background to an animal of preferable genetic background. In gene knockout experiments in particular, where the knockout is performed on easily cultured stem cell lines, but is required in an animal with a different genetic background, the knockout animal is backcrossed against the animal of the required genetic background. As the figure shows, each time that the mouse with the desired trait (in this case the lack of a gene (i.e. a knockout), indicated by the presence of a positive selectable marker) is crossed with a mouse of a constant genetic background, the average percentage of the genetic material of the offspring that is derived from that constant background increases. The result, after sufficient reiterations, is an animal with the desired trait in the desired genetic background, with the percentage of genetic material from the original stem cells reduced to a minimum (in the order of 0.01%).

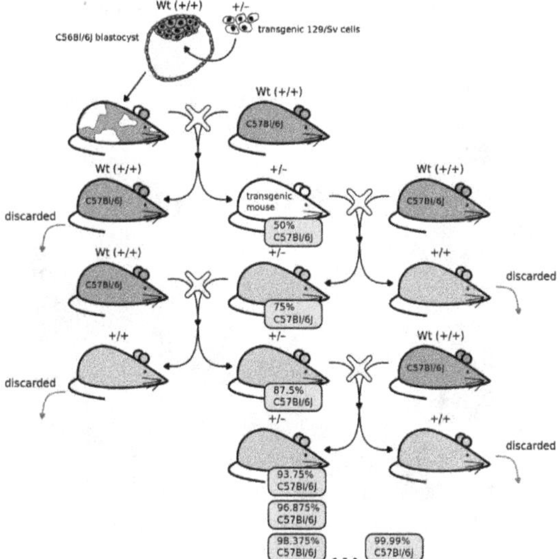

This image demonstrates backcrossing of a heterozygous mouse from one genetic background onto another genetic background. In this example, the gene knockout is performed on 129/Sv cells and then backcrossed into the C57B/6J genetic background. With each successive backcross, there is an increase in the percentage of C57B/6J DNA that constitutes the genome of the offspring.

Due to the nature of meiosis, in which chromosomes derived from each parent are randomly shuffled and assigned to each nascent gamete, the percentage of genetic material deriving from either cell-line will vary between offspring of a single crossing but will have an expected value. The genotype of each member of offspring may be assessed to choose not only an individual that carries the desired genetic trait, but also the minimum percentage of genetic material from the original stem cell line.

A *consomic strain* is an inbred strain with one of its chromosomes replaced by the homologous chromosome of another inbred strain via a series of marker-assisted backcrosses.

Breeding Back

Breeding back is a form of artificial selection by the deliberate selective breeding of domestic animals, in an attempt to achieve an animal breed with a phenotype that resembles a wildtype ancestor, usually one that is extinct.

Heck cattle have been bred to resemble the aurochs in the 1920s. Since they are regarded as not meeting the criteria for "new aurochs", recent efforts such as the TaurOs Project and Uruz Project attempt to get considerably closer to the wild aurochs.

Heck horse in Haselünne, Germany.

It must be kept in mind that a breeding-back breed may be very similar to the extinct wild type in phenotype, ecological niche, and to some extent genetics, but the initial gene pool of that wild type is lost forever with its extinction. It is not truly possible for a breeding back attempt to actually recreate an extinct wild type that is the breeding target, as an extinct wild type cannot be resurrected via selective breeding alone. Furthermore, even the outward authenticity of a bred-back type depends on the quality of the stock used as the foundation of the project. As a result, some breeds, like Heck cattle, are at best a vague look-alike of the extinct wildtype aurochs, according to the literature.

Background

It is the target of breeding back to restore the wild traits that may have been preserved in domestic animals within one breeding lineage. Commonly, not only the phenotype but also the ecological capacity are considered in breeding back projects, because hardy breeding back results are used in certain conservation projects. In nature, usually only individuals suited to the natural circumstances can survive and reproduce, while humans select those with additional attractive, docile or productive characteristics. Therefore, selection criteria in nature versus those in domestic conditions are different and domesticated animals often differ significantly in phenotype, behaviour and genetics from their wild forerunners. It is the target of breeding back to re-create the wild traits that may have been preserved in domestic animals within one breeding lineage.

In many cases, the extinct wild type ancestors of a given species are known only through skeletons and, in some cases, historical descriptions, so their phenotype is not directly accessible. Therefore, there is no certainty of success with a breeding back attempt and the results must be reviewed with caution. In order to test genetic closeness, both mitochondrial DNA and nuclear DNA have to be accessible. However, success is possible: humans selected only for superficial characters and as a rule did not change inner mechanisms such as digestion. Further, since many domestic animals show a behaviour that is derived from their wild ancestors (such as the herding instinct of cattle or the social instincts of dogs), and are fit to survive under natural circumstances, as evidenced by the many feral populations of many domestic animals, it can be presumed that back-bred animals will function like their wild ancestors. For example, food choice should be the same in domestic and wild types. Natural selection could serve as an additional tool in creating authentic robustness, behaviour and maybe also phenotype. For large herbivores, a sufficient predator population would also be necessary to support such a process, but this predator population is largely not present in today's Europe, where many breeding back projects are conducted.

There is a difference between creating a look-alike breed from domestic animals and creating a look-alike via artificial selection of individuals of a different species or subspecies that may resemble the animal that is targeted. For example, the Quagga was an independent subspecies of the plains zebra, and has no descendants which preserved its features. Thus, the Quagga Project tries to breed a look-alike out of living plains zebras which have fewer stripes than average, selecting for reduced striping. The resulting animal will bear only a superficial resemblance to the extinct Quagga.

Use

Bred-back breeds are interesting for conservation biology since they may fill an ecological gap that has been left due to the extinction of the wildtypes due to hunting by humans. As long as food choice, behaviour, robustness, defence against predators, hunting or foraging instincts and phenotype are the same as in the wild type, it will function like it in the wilderness. Releasing such animals into the wild would fill the previously empty niche and allows a natural dynamic between the species of the ecosystem to re-evolve. However, breeding-back attempts do not always result in an animal that is closer to the wild type than other domestic breeds. For example, Heck cattle bear less resemblance to the aurochs than many Iberian fighting cattle do.

Examples

Aurochs

Ideas for creating an aurochs-like animal from domestic cattle have been around since 1835. In the 1920s, Heinz and Lutz Heck tried to breed a look-alike using central European dairy breeds and southern European cattle. The result, Heck cattle, are hardy, but differ from the aurochs in many respects, although a resemblance in colour and sometimes also horns has been achieved. From 1996 onwards, Heck cattle have been crossed with primitive Iberian breeds like Sayaguesa Cattle and Fighting cattle, and the very large Italian breed Chianina in a number of German reserves in order to enhance the resemblance to the aurochs. The results are called Taurus cattle, which are larger and more long-legged than Heck cattle and have a more aurochs-like horn shape. Another project to achieve a type of cattle that resembles the aurochs is TaurOs Project, using primitive and hardy southern European breeds and also Scottish Highland cattle. These Tauros cattle are planned to be established in various European natural reserves.

European Wild Horse

The Polish Konik horse is often erroneously considered the result of a breeding-back experiment to "recreate" the phenotype of the Tarpan, the European wild horse. The Konik is actually a hardy landrace breed originating in Poland which was called *Panje horse* before agriculturist Tadeusz Vetulani coined the name "Konik" in the 1920s. Vetulani did start an experiment trying to reconstruct the Tarpan using Koniks, but his stock only had a minor contribution to the modern Konik population.

However, during World War II, the Heck brothers crossed Koniks with Przewalski's horses and also other ponies like Icelandic horses and Gotland ponies; the result is now called the Heck Horse. During the last decades, Heck horses have continuously been crossed with Koniks, so that their phenotype now is nearly indistinguishable, except that the Heck horse has a slenderer build on average.

Besides the Konik-type horses, the Exmoor Pony that still lives in a semi-feral state in some parts of Southern England has also been claimed to be closer to the European wild horse as it carries pangare coloration and has a resemblance to the horses in Lascaux cave paintings.

Wolf

Although the wolf, the wild ancestor of domestic dogs, is not extinct, its phenotype is the target of several developmental breeds including the Northern Inuit Dog and the Tamaskan Dog. They are all crossbreeds of German Shepherd Dogs and huskies, selected for phenotypic wolf characteristics. Therefore, these breeds are breeding-back attempts as well.

A Tamaskan Dog, bred to visually resemble a wolf.

Culling

In biology, culling is the process of segregating organisms from a group according to desired or undesired characteristics. In animal breeding, culling is the process of removing or segregating animals from a breeding stock based on specific criteria. This is done either to reinforce or exaggerate desirable characteristics, or to remove undesirable characteristics from the group. For livestock and wildlife, culling often refers to the act of killing removed animals. In fruits and vegetables, culling is the sorting or segregation of fresh harvested produce into marketable lots, with the non-marketable lots being discarded or diverted into food processing or non-food processing activities. This usually happens at collection centres located at, or close to farms. Culling is sometimes used as a term to describe indiscriminate killing within one particular species which can be due to a range of reasons, for example, badger culling in the United Kingdom.

Drafting out culled sheep

Origin of the Term

The word comes from the Latin *colligere*, which means "to collect". The term can be applied broadly to mean sorting a collection into two groups: one that will be kept and one that will be rejected. The cull is the set of items rejected during the selection process. The culling process is repeated until the selected group is of proper size and consistency desired.

Pedigreed Animals

Culling The rejection or removal of inferior individuals from breeding. The act of selective breeding. As used in the practice of breeding pedigree cats, this refers to the practice of spaying or neutering a kitten or cat that does not measure up to the show standard (or other standard being applied) for that breed. In no way does culling, as used by responsible breeders, signify the killing of healthy kittens or cats if they fail to meet the applicable standard."

Robinson's Genetics for Cat Breeders and Veterinarians, Fourth Edition

In the breeding of pedigreed animals, both desirable and undesirable traits are considered when choosing which animals to retain for breeding and which to place as pets. The process of culling starts with examination of the conformation standard of the animal and will often include additional qualities such as health, robustness, temperament, color preference, etc. The breeder takes all things into consideration when envisioning his/her ideal for the breed or goal of their breeding program. From that vision, selections are made as to which animals, when bred, have the best chance of producing the ideal for the breed.

Breeders of pedigreed animals cull based on many criteria. The first culling criterion should always be health and robustness. Secondary to health, temperament and conformation of the animal should be considered. The filtering process ends with the breeder's personal preferences on pattern, color, etc.

The Tandem Method

The Tandem Method is a form of selective breeding where a breeder addresses one characteristic of the animal at a time, thus selecting only animals that measure above a certain threshold for that particular trait while keeping other traits constant. Once that level of quality in the single trait is achieved, the breeder will focus on a second trait and cull based on that quality. With the tandem method, a minimum level of quality is set for important characteristics that the breeder wishes to remain constant. The breeder is focussing improvement in one particular trait without losing quality of the others. The breeder will raise the threshold for selection on this trait with each successive generation of progeny, thus ensuring improvement in this single characteristic of his breeding program.

For example, a breeder that is pleased with the muzzle length, muzzle shape, and eye placement in the breeding stock, but wishes to improve the eye shape of progeny produced may determine a minimum level of improvement in eye shape required for progeny to be returned into the breeding program. Progeny is first evaluated on the existing quality thresholds in place for muzzle length, muzzle shape, and eye placement with the additional criterion being improvement in eye shape. Any animal that does not meet this level of improvement in the eye shape while maintaining the other qualities is culled from the breeding program; i.e., that animal is not used for breeding, but is instead spayed/neutered and placed in a pet home.

Independent Levels

Independent levels is a method where any animal who falls below a given standard in any single characteristic is not used in a breeding program. With each successive mating, the threshold culling criteria is raised thus improving the breed with each successive generation.

This method measures several characteristics at once. Should progeny fall below the desired quality in any one characteristic being measured, it will not be used in the breeding program regardless of the level of excellence of other traits. With each successive generation of progeny, the minimum quality of each characteristic is raised thus insuring improvement of these traits.

For example, a breeder has a view of what the minimum requirements for muzzle length, muzzle shape, eye placement, and eye shape she is breeding toward. The breeder will determine what the minimum acceptable quality for each of these traits will be for progeny to be folded back into her breeding program. Any animal that fails to meet the quality threshold for any one of these criteria is culled from the breeding program.

Total Score Method

The Total Score Method is a method where the breeder evaluates and selects breeding stock based on a weighted table of characteristics. The breeder selects qualities that are most important to them and assigns them a weight. The weights of all the traits should add up to 100. When evaluating an individual for selection, the breeder measures the traits on a scale of 1 to 10, with 10 being the most desirable expression and 1 being the lowest. The scores are then multiplied by their weights and then added together to give a total score. Individuals that fail to meet a threshold are culled (or removed) from the breeding program. The total score gives a breeder a way to evaluate multiple traits on an animal at the same time.

The total score method is the most flexible of the three. it allows for weighted improvement of multiple characteristics. It allows the breeder to make major gains in one aspect while moderate or lesser gains in others.

For example, a breeder is willing to make a smaller improvement in muzzle length and muzzle shape in order to have a moderate gain in improvement of eye placement and a more dramatic improvement in eye shape. Suppose the breeder determines that she would like to see 40% improvement in eye shape, 30% improvement in eye placement, and 15% improvement in both muzzle length and shape. The breeder would evaluate these characteristics on a scale of 1 to 10 and multiply by the weights. The formula would look something like: 15 (muzzle length) + 15(muzzle shape) + 30(eye placement) + 40(eye shape) = total score for that animal. The breeder determines the lowest acceptable total score for an animal to be folded back into their breeding program. Animals that do not meet this minimum total score are culled from the breeding program.

Livestock and Production Animals

Since livestock is bred for the production of meat or milk, the herd must be culled to a certain number of production or meat animals a farmer wishes to maintain. Animals not selected to remain for breeding are sent to the slaughter house, sold, or killed.

Criteria for culling livestock and production animals can be based on population or production (milk or egg). In a domestic or farming situation the culling process involves selection and the selling of surplus stock. The selection may be done to improve breeding stock, for example for improved production of eggs or milk, or simply to control the group's population for the benefit of the environment and other species.

With dairy cattle, culling may be practised by inseminating inferior cows with beef breed semen and by selling the produced offspring for meat production.

With poultry, males which would grow up to be roosters have little use in an industrial egg-producing facility. Approximately half of the newly hatched chicks will be male and would grow up to be roosters, which do not lay eggs. For this reason, the hatchlings are culled based on gender. Most of the male chicks are usually killed shortly after hatching.

Wildlife

In the United States, hunting licenses and hunting seasons are a means by which the population of game animals is maintained. Each season, a hunter is allowed to kill a certain amount of wild game. The amount is determined both by species and gender. If the population seems to have surplus females, hunters are allowed to take more females during that hunting season. If the population is below what is desired, hunters may not be permitted to hunt that particular game animal or only hunt a restricted number of males.

Populations of game animals such as elk may be informally culled if they begin to excessively eat winter food set out for domestic cattle herds. In such instances the rancher will inform hunters that they may "hunt the haystack" on his property in order to thin the wild herd to controllable levels. These efforts are aimed to counter excessive depletion of the intended "domestic" winter feed supplies. Other managed culling instances involve extended issuance of extra hunting licenses, or the inclusion of additional "special hunting seasons" during harsh winters or overpopulation periods, governed by state fish and game Agencies.

Culling for population control is common in wildlife management, particularly on African game farms and in Australia in national parks. In the case of very large animals such as elephants, adults are often targeted. Their orphaned young, easily captured and transported, are then relocated. Without proper elephant socialization, young male elephants are believed to become unruly and extremely dangerous to other elephants, wildlife and humans. Culling is controversial in many African countries, but reintroduction of the practice has been recommended in recent years for use at the Kruger National Park in South Africa, which has experienced a swell in its elephant population since culling was banned in 1995.

Birds

Some bird species are culled when their populations impact upon human property, business or recreational activity, disturb or modify habitats or otherwise impact species of conservation concern. Cormorants are culled in many countries due to their impact on commercial and recreational fisheries and habitat modification for nesting and guano deposition. They are culled by shooting and the smothering of eggs with oil. Another example is the culling of silver gulls in order to protect the chicks of the vulnerable banded stilt at ephemeral inland salt lake breeding sites in South Australia. The gulls were culled using bread laced with a narcotic substance. In the Australian states of Tasmania and South Australia, Cape Barren geese are culled to limit damage to crops and the fouling of waterholes. Cape Barren Geese remain one of the rarest geese in the world, though much of their habitat is now regarded as secure.

Double crested cormorant

Seals

In South Australia, the recovery of the state's native population of New Zealand fur seals (*Arctocephalus forsteri*) after severe depletion by sealers in the 1800s has brought them into conflict with the fishing industry. This has prompted members of Parliament to call for seal culling in South Australia. The State Government continues to resist the pressure and as of July 2015, the animals remain protected as listed Marine Mammals under the state's *National Parks and Wildlife Act 1972*.

New Zealand fur seal

Sharks

In 2014, a controversial policy was introduced by the Western Australian state government which became known as the Western Australian shark cull. Baited hooks known as drum lines were to be set over several consecutive summers in order to catch and kill otherwise protected great white sharks. The policy's objective was to protect users of the marine environment from fatal shark attack. Thousands of people protested against its implementation, claiming that it was indiscriminate, inhumane and worked against scientific advice the government had previously received. Seasonal setting of drum lines was abandoned in September 2014 after the program failed to catch any great white sharks, instead catching 172 other elasmobranchii, mostly tiger sharks.

Great white shark

Culling in Zoos

Many zoos participate in an international breeding program to maintain a genetically-viable population and prevent inbreeding. Animals that can no longer contribute to the breeding program are considered less desirable and are often replaced by more desirable individuals. If an animal is surplus to a zoo's requirements and a place in another zoo can not be found, the animal may be killed. In 2014, the culling of a young, healthy giraffe Marius raised an international public controversy.

Zoos sometimes consider female animals to be more desirable than males. One reason for this is that while individual males can contribute to the birth of many young in a short period of time, females give birth to only a few young and are pregnant for a relatively long period of time. This makes it possible to keep many females with just one or two males, but not the reverse. Another reason is that the birth of some animal species increases public interest in the zoo.

Animal law of Germany (Tierschutzgesetz (de)) orders that zoo animals cannot be culled without verification by official veterinary institutes of the Landkreis or federated state. In the U.K., there is no general prohibition on euthanasia and this is allowed when over-crowding compromises the well-being of the animals.

Culling and Ethics

Jaak Panksepp, one of America's leading neuroscientists, concludes that both animals and humans have brains wired to feel emotions, and that animals have the capacity to experience pleasure and happiness from their lives. For this reason, there are those who believe that culling animals is morally wrong.

Some argue that culling is necessary when biodiversity is threatened. However, the protection of biodiversity argument has been questioned by some animal rights advocates who point out that the animal which most greatly threatens and damages biodiversity is humanity, so if we are not willing to cull our own species we cannot morally justify culling another.

References

- Jain, H. K.; Kharkwal, M. C. (2004). Plant breeding - Mendelian to molecular approaches. Boston, London, Dordecht: Kluwer Academic Publishers. ISBN 1-4020-1981-5.
- Gjedrem, T & Baranski, M. (2009). Selective breeding in Aquaculture: An Introduction. 1st Edition. Springer. ISBN 978-90-481-2772-6
- Buffum, Burt C. (2008). Arid Agriculture; A Hand-Book for the Western Farmer and Stockman. Read Books. p. 232. ISBN 978-1-4086-6710-1.
- Grandin, Temple; Johnson, Catherine (2005). Animals in Translation. New York, New York: Scribner. pp. 69–71. ISBN 0-7432-4769-8.
- Thornhill, Nancy Wilmsen (1993). The Natural History of Inbreeding and Outbreeding: Theoretical and Empirical Perspectives. Chicago: University of Chicago Press. ISBN 0-226-79854-2.
- Shields, William M. (1982). Philopatry, inbreeding, and the evolution of sex. Albany: State University of New York Press. ISBN 0-87395-618-4.
- Hartl, D.L., Jones, E.W. (2000) Genetics: Analysis of Genes and Genomes. Fifth Edition. Jones and Bartlett Publishers Inc., pp. 105–106. ISBN 0-7637-1511-5.
- Kingston H M (2002). ABC of Clinical Genetics (3rd ed.). London: BMJ Books. p. 7. ISBN 0-7279-1627-0. PMC 1836181. PMID 2497870.
- Wolf, Arthur P.; Durham, William H., eds. (2005). Inbreeding, incest, and the incest taboo: the state of knowledge at the turn. Stanford University Press. p. 3. ISBN 0-8047-5141-2.
- Griffiths, Anthony J. F.; Jeffrey H. Miller; David T. Suzuki; Richard C. Lewontin; William M. Gelbart (1999). An introduction to genetic analysis. New York: W. H. Freeman. pp. 726–727. ISBN 0-7167-3771-X.
- Wolf, Arthur P.; Durham, William H., eds. (2005). Inbreeding, incest, and the incest taboo: the state of knowledge at the turn. Stanford University Press. p. 6. ISBN 0-8047-5141-2.
- Tave, Douglas (1999). Inbreeding and brood stock management. Food and Agriculture Organization of the United Nations. p. 50. ISBN 978-92-5-104340-0.
- Beeche, Arturo (2009). The Gotha: Still a Continental Royal Family, Vol. 1. Richmond, US: Kensington House Books. pp. 1–13. ISBN 978-0-9771961-7-3.
- Cis van Vuure: Retracing the Aurochs – History, Morphology and Ecology of an extinct wild Ox. 2005. ISBN 954-642-235-5
- Tadeusz Jezierski, Zbigniew Jaworski: Das Polnische Konik. Die Neue Brehm-Bücherei Bd. 658, Westarp Wissenschaften, Hohenwarsleben 2008, ISBN 3-89432-913-0
- Robinson, Roy; Carolyn M. Vella; Lorraine M. Shelton; John J. McGonagle; Terry W. Stanglein (1999). Robinson's Genetics for Cat Breeders and Veterinarians (Fourth ed.). Great Britain: Butterworth Heinemann. ISBN 0-7506-4069-3.

4

Gene: A Comprehensive Study

A gene is the functional unit of heredity. It is very important to understand gene nomenclature, gene expression, essential gene and heredity, to have a profound understanding of genes. This chapter has been carefully written to provide an easy understanding of the varied facets of genes.

Gene

A gene is a locus (or region) of DNA which is made up of nucleotides and is the molecular unit of heredity. The transmission of genes to an organism's offspring is the basis of the inheritance of phenotypic traits. Most biological traits are under the influence of polygenes (many different genes) as well as the gene–environment interactions. Some genetic traits are instantly visible, such as eye colour or number of limbs, and some are not, such as blood type, risk for specific diseases, or the thousands of basic biochemical processes that comprise life.

Genes can acquire mutations in their sequence, leading to different variants, known as alleles, in the population. These alleles encode slightly different versions of a protein, which cause different phenotype traits. Colloquial usage of the term "having a gene" (e.g., "good genes," "hair colour gene") typically refers to having a different allele of the gene. Genes evolve due to natural selection or survival of the fittest of the alleles.

The concept of a gene continues to be refined as new phenomena are discovered. For example, regulatory regions of a gene can be far removed from its coding regions, and coding regions can be split into several exons. Some viruses store their genome in RNA instead of DNA and some gene products are functional non-coding RNAs. Therefore, a broad, modern working definition of a gene is any discrete locus of heritable, genomic sequence which affect an organism's traits by being expressed as a functional product or by regulation of gene expression.

Discovery of Discrete Inherited Units

The existence of discrete inheritable units was first suggested by Gregor Mendel (1822–1884). From 1857 to 1864, he studied inheritance patterns in 8000 common edible pea plants, tracking distinct traits from parent to offspring. He described these mathematically as 2^n combinations where n is the number of differing characteristics in the original peas. Although he did not use the term *gene*, he explained his results in terms of discrete inherited units that give rise to observable physical characteristics. This description prefigured the distinction between genotype (the genetic material of an organism) and phenotype (the visible traits of that organism). Mendel was also the first to demonstrate independent assortment, the distinction between dominant and recessive traits, the distinction between a heterozygote and homozygote, and the phenomenon of discontinuous inheritance.

Prior to Mendel's work, the dominant theory of heredity was one of blending inheritance, which suggested that each parent contributed fluids to the fertilisation process and that the traits of the parents blended and mixed to produce the offspring. Charles Darwin developed a theory of inheritance he termed pangenesis, from Greek pan ("all, whole") and genesis ("birth") / genos ("origin"). Darwin used the term *gemmule* to describe hypothetical particles that would mix during reproduction.

Mendel's work went largely unnoticed after its first publication in 1866, but was rediscovered in the late 19th century by Hugo de Vries, Carl Correns, and Erich von Tschermak, who (claimed to have) reached similar conclusions in their own research. Specifically, in 1889, Hugo de Vries published his book *Intracellular Pangenesis*, in which he postulated that different characters have individual hereditary carriers and that inheritance of specific traits in organisms comes in particles. De Vries called these units "pangenes" (*Pangens* in German), after Darwin's 1868 pangenesis theory.

Sixteen years later, in 1905, the word *genetics* was first used by William Bateson, while Eduard Strasburger, amongst others, still used the term pangene for the fundamental physical and functional unit of heredity. In 1909 the Danish botanist Wilhelm Johannsen shortened the name to "gene".

Discovery of DNA

Advances in understanding genes and inheritance continued throughout the 20th century. Deoxyribonucleic acid (DNA) was shown to be the molecular repository of genetic information by experiments in the 1940s to 1950s. The structure of DNA was studied by Rosalind Franklin and Maurice Wilkins using X-ray crystallography, which led James D. Watson and Francis Crick to publish a model of the double-stranded DNA molecule whose paired nucleotide bases indicated a compelling hypothesis for the mechanism of genetic replication.

In the early 1950s the prevailing view was that the genes in a chromosome acted like discrete entities, indivisible by recombination and arranged like beads on a string. The experiments of Benzer using mutants defective in the rII region of bacteriophage T4 (1955-1959) showed that individual genes have a simple linear structure and are likely to be equivalent to a linear section of DNA.

Collectively, this body of research established the central dogma of molecular biology, which states that proteins are translated from RNA, which is transcribed from DNA. This dogma has since been shown to have exceptions, such as reverse transcription in retroviruses. The modern study of genetics at the level of DNA is known as molecular genetics.

In 1972, Walter Fiers and his team at the University of Ghent were the first to determine the sequence of a gene: the gene for Bacteriophage MS2 coat protein. The subsequent development of chain-termination DNA sequencing in 1977 by Frederick Sanger improved the efficiency of sequencing and turned it into a routine laboratory tool. An automated version of the Sanger method was used in early phases of the Human Genome Project.

Modern Evolutionary Synthesis

The theories developed in the 1930s and 1940s to integrate molecular genetics with Darwinian evolution are called the modern evolutionary synthesis, a term introduced by Julian Huxley.

Evolutionary biologists subsequently refined this concept, such as George C. Williams' gene-centric view of evolution. He proposed an evolutionary concept of the gene as a unit of natural selection with the definition: "that which segregates and recombines with appreciable frequency." In this view, the molecular gene *transcribes* as a unit, and the evolutionary gene *inherits* as a unit. Related ideas emphasizing the centrality of genes in evolution were popularized by Richard Dawkins.

Molecular Basis

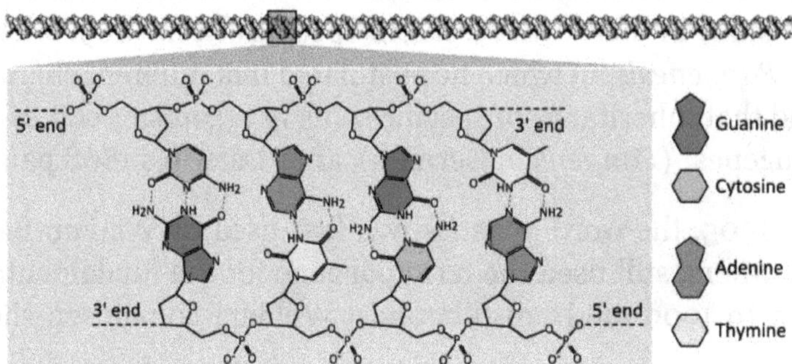

The chemical structure of a four base pair fragment of a DNA double helix. The sugar-phosphate backbone chains run in opposite directions with the bases pointing inwards, base-pairing A to T and C to G with hydrogen bonds.

DNA

The vast majority of living organisms encode their genes in long strands of DNA (deoxyribonucleic acid). DNA consists of a chain made from four types of nucleotide subunits, each composed of: a five-carbon sugar (2'-deoxyribose), a phosphate group, and one of the four bases adenine, cytosine, guanine, and thymine.

Two chains of DNA twist around each other to form a DNA double helix with the phosphate-sugar backbone spiralling around the outside, and the bases pointing inwards with adenine base pairing to thymine and guanine to cytosine. The specificity of base pairing occurs because adenine and thymine align to form two hydrogen bonds, whereas cytosine and guanine form three hydrogen bonds. The two strands in a double helix must therefore be complementary, with their sequence of bases matching such that the adenines of one strand are paired with the thymines of the other strand, and so on.

Due to the chemical composition of the pentose residues of the bases, DNA strands have directionality. One end of a DNA polymer contains an exposed hydroxyl group on the deoxyribose; this is known as the 3' end of the molecule. The other end contains an exposed phosphate group; this is the 5' end. The two strands of a double-helix run in opposite directions. Nucleic acid synthesis, including DNA replication and transcription occurs in the 5'→3' direction, because new nucleotides are added via a dehydration reaction that uses the exposed 3' hydroxyl as a nucleophile.

The expression of genes encoded in DNA begins by transcribing the gene into RNA, a second type of nucleic acid that is very similar to DNA, but whose monomers contain the sugar ribose rather than deoxyribose. RNA also contains the base uracil in place of thymine. RNA molecules are less stable than DNA and are typically single-stranded. Genes that encode proteins are composed of a series of three-nucleotide sequences called codons, which serve as the "words" in the genetic "language". The genetic code specifies the correspondence during protein translation between codons and amino acids. The genetic code is nearly the same for all known organisms.

Chromosomes

The total complement of genes in an organism or cell is known as its genome, which may be stored on one or more chromosomes. A chromosome consists of a single, very long DNA helix on which thousands of genes are encoded. The region of the chromosome at which a particular gene is located is called its locus. Each locus contains one allele of a gene; however, members of a population may have different alleles at the locus, each with a slightly different gene sequence.

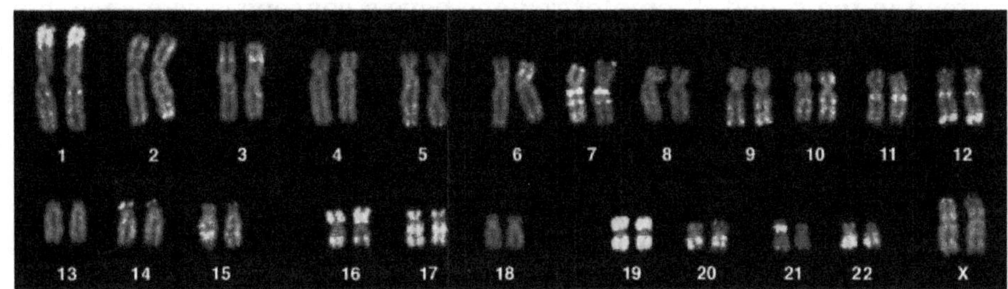

Fluorescent microscopy image of a human female karyotype, showing 23 pairs of chromosomes . The DNA is stained red, with regions rich in housekeeping genes further stained in green. The largest chromosomes are around 10 times the size of the smallest.

The majority of eukaryotic genes are stored on a set of large, linear chromosomes. The chromosomes are packed within the nucleus in complex with storage proteins called histones to form a unit called a nucleosome. DNA packaged and condensed in this way is called chromatin. The manner in which DNA is stored on the histones, as well as chemical modifications of the histone itself, regulate whether a particular region of DNA is accessible for gene expression. In addition to genes, eukaryotic chromosomes contain sequences involved in ensuring that the DNA is copied without degradation of end regions and sorted into daughter cells during cell division: replication origins, telomeres and the centromere. Replication origins are the sequence regions where DNA replication is initiated to make two copies of the chromosome. Telomeres are long stretches of repetitive sequence that cap the ends of the linear chromosomes and prevent degradation of coding and regulatory regions during DNA replication. The length of the telomeres decreases each time the genome is replicated and has been implicated in the aging process. The centromere is required for binding spindle fibres to separate sister chromatids into daughter cells during cell division.

Prokaryotes (bacteria and archaea) typically store their genomes on a single large, circular chromosome. Similarly, some eukaryotic organelles contain a remnant circular chromosome with a small number of genes. Prokaryotes sometimes supplement their chromosome with additional small circles of DNA called plasmids, which usually encode only a few genes and are transferable between individuals. For example, the genes for antibiotic resistance are usually encoded on bacterial plasmids and can be passed between individual cells, even those of different species, via horizontal gene transfer.

Whereas the chromosomes of prokaryotes are relatively gene-dense, those of eukaryotes often contain regions of DNA that serve no obvious function. Simple single-celled eukaryotes have relatively small amounts of such DNA, whereas the genomes of complex multicellular organisms, including humans, contain an absolute majority of DNA without an identified function. This DNA has often been referred to as "junk DNA". However, more recent analyses suggest that, although protein-coding DNA makes up barely 2% of the human genome, about 80% of the bases in the genome may be expressed, so the term "junk DNA" may be a misnomer.

Structure and Function

Firstly, flanking the open reading frame, all genes contain a regulatory sequence that is required for their expression. In order to be expressed, genes require a promoter sequence. The promoter is recognized and bound by transcription factors and RNA polymerase to initiate transcription. A gene can have more than one promoter, resulting in messenger RNAs (mRNA) that differ in how far they extend in the 5' end. Promoter regions have a consensus sequence, however highly transcribed genes have "strong" promoter sequences that bind the transcription machinery well, whereas others have "weak" promoters that bind poorly and initiate transcription less frequently. Eukaryotic promoter regions are much more complex and difficult to identify than prokaryotic promoters.

Additionally, genes can have regulatory regions many kilobases upstream or downstream of the open reading frame. These act by binding to transcription factors which then cause the DNA to loop so that the regulatory sequence (and bound transcription factor) become close to the RNA polymerase binding site. For example, enhancers increase transcription by binding an activator protein which then helps to recruit the RNA polymerase to the promoter; conversely silencers bind repressor proteins and make the DNA less available for RNA polymerase.

The transcribed pre-mRNA contains untranslated regions at both ends which contain a ribosome binding site, terminator and start and stop codons. In addition, most eukaryotic open reading frames contain untranslated introns which are removed before the exons are translated. The sequences at the ends of the introns, dictate the splice sites to generate the final mature mRNA which encodes the protein or RNA product.

Many prokaryotic genes are organized into operons, with multiple protein-coding sequences that are transcribed as a unit. The genes in an operon are transcribed as a continuous messenger RNA, referred to as a polycistronic mRNA. The term cistron in this context is equivalent to gene. The transcription of an operon's mRNA is often controlled by a repressor that can occur in an active or inactive state depending on the presence of certain specific metabolites. When active, the repressor binds to a DNA sequence at the beginning of the operon, called the operator region, and represses transcription of the operon; when the repressor is inactive transcription of the operon can occur (e.g. Lac operon). The products of operon genes typically have related functions and are involved in the same regulatory network.

Functional Definitions

Defining exactly what section of a DNA sequence comprises a gene is difficult. Regulatory regions of a gene such as enhancers do not necessarily have to be close to the coding sequence on the linear molecule because the intervening DNA can be looped out to bring the gene and its regulatory region into proximity. Similarly, a gene's introns can be much larger than its exons. Regulatory regions can even be on entirely different chromosomes and operate *in trans* to allow regulatory regions on one chromosome to come in contact with target genes on another chromosome.

Early work in molecular genetics suggested the concept that one gene makes one protein. This concept (originally called the one gene-one enzyme hypothesis) emerged from an influential 1941 paper by George Beadle and Edward Tatum on experiments with mutants of the fungus Neurospora crassa. Norman Horowitz, an early colleague on the *Neurospora* research, reminisced in 2004 that "these experiments founded the science of what Beadle and Tatum called *biochemical genetics*. In actuality they proved to be the opening gun in what became molecular genetics and all the developments that have followed from that." The one gene-one protein concept has been refined since the discovery of genes that can encode multiple proteins by alternative splicing and coding sequences split in short section across the genome whose mRNAs are concatenated by trans-splicing.

A broad operational definition is sometimes used to encompass the complexity of these diverse phenomena, where a gene is defined as a union of genomic sequences encoding a coherent set of potentially overlapping functional products. This definition categorizes genes by their functional products (proteins or RNA) rather than their specific DNA loci, with regulatory elements classified as *gene-associated* regions.

Gene Expression

In all organisms, two steps are required to read the information encoded in a gene's DNA and produce the protein it specifies. First, the gene's DNA is *transcribed* to messenger RNA (mRNA). Second, that mRNA is *translated* to protein. RNA-coding genes must still go through the first step, but are not translated into protein. The process of producing a biologically functional molecule of either RNA or protein is called gene expression, and the resulting molecule is called a gene product.

Genetic Code

The nucleotide sequence of a gene's DNA specifies the amino acid sequence of a protein through the genetic code. Sets of three nucleotides, known as codons, each correspond to a specific amino acid. The principle that three sequential bases of DNA code for each amino acid was demonstrated in 1961 using frameshift mutations in the rIIB gene of bacteriophage T4.

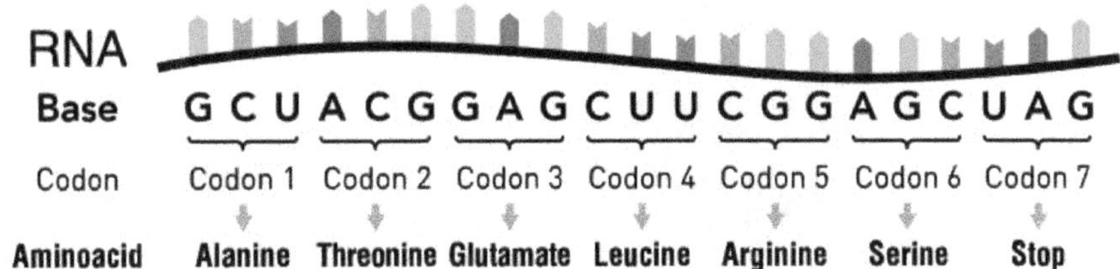

Schematic of a single-stranded RNA molecule illustrating a series of three-base codons. Each three-nucleotide codon corresponds to an amino acid when translated to protein

Additionally, a "start codon", and three "stop codons" indicate the beginning and end of the protein coding region. There are 64 possible codons (four possible nucleotides at each of three positions, hence 4^3 possible codons) and only 20 standard amino acids; hence the code is redundant and multiple codons can specify the same amino acid. The correspondence between codons and amino acids is nearly universal among all known living organisms.

Transcription

Transcription produces a single-stranded RNA molecule known as messenger RNA, whose nucleotide sequence is complementary to the DNA from which it was transcribed. The mRNA acts as an intermediate between the DNA gene and its final protein product. The gene's DNA is used as a template to generate a complementary mRNA. The mRNA matches the sequence of the gene's DNA coding strand because it is synthesised as the complement of the template strand. Transcription is performed by an enzyme called an RNA polymerase, which reads the template strand in the 3' to 5' direction and synthesizes the RNA from 5' to 3'. To initiate transcription, the polymerase first recognizes and binds a promoter region of the gene. Thus, a major mechanism of gene regulation is the blocking or sequestering the promoter region, either by tight binding by repressor molecules that physically block the polymerase, or by organizing the DNA so that the promoter region is not accessible.

In prokaryotes, transcription occurs in the cytoplasm; for very long transcripts, translation may begin at the 5' end of the RNA while the 3' end is still being transcribed. In eukaryotes, transcription occurs in the nucleus, where the cell's DNA is stored. The RNA molecule produced by the polymerase is known as the primary transcript and undergoes post-transcriptional modifications before being exported to the cytoplasm for translation. One of the modifications performed is the splicing of introns which are sequences in the transcribed region that do not encode protein. Alternative splicing mechanisms can result in mature transcripts from the same gene having different sequences and thus coding for different proteins. This is a major form of regulation in eukaryotic cells and also occurs in some prokaryotes.

Translation

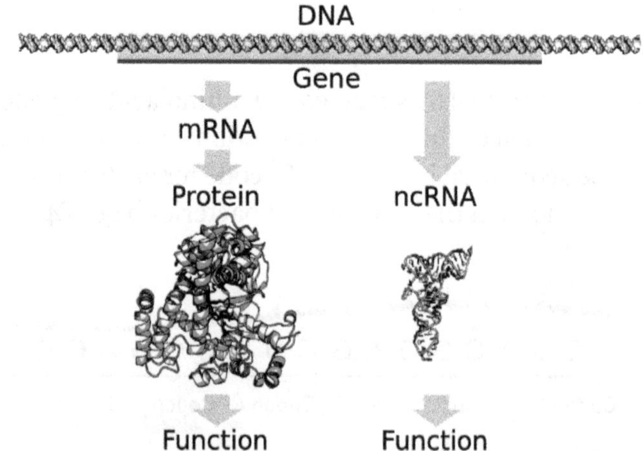

Protein coding genes are transcribed to an mRNA intermediate, then translated to a functional protein. RNA-coding genes are transcribed to a functional non-coding RNA. (PDB: 3BSE, 1OBB, 3TRA)

Translation is the process by which a mature mRNA molecule is used as a template for synthesizing a new protein. Translation is carried out by ribosomes, large complexes of RNA and protein responsible for carrying out the chemical reactions to add new amino acids to a growing polypeptide chain by the formation of peptide bonds. The genetic code is read three nucleotides at a time, in units called codons, via interactions with specialized RNA molecules called transfer RNA (tRNA). Each tRNA has three unpaired bases known as the anticodon that are complementary to the codon it reads on the mRNA. The tRNA is also covalently attached to the amino acid specified by the complementary codon. When the tRNA binds to its complementary codon in an mRNA strand, the ribosome attaches its amino acid cargo to the new polypeptide chain, which is synthesized from amino terminus to carboxyl terminus. During and after synthesis, most new proteins must fold to their active three-dimensional structure before they can carry out their cellular functions.

Regulation

Genes are regulated so that they are expressed only when the product is needed, since expression draws on limited resources. A cell regulates its gene expression depending on its external environment (e.g. available nutrients, temperature and other stresses), its internal environment (e.g. cell division cycle, metabolism, infection status), and its specific role if in a multicellular organism. Gene expression can be regulated at any step: from transcriptional initiation, to RNA processing, to post-translational modification of the protein. The regulation of lactose metabolism genes in *E. coli* (*lac* operon) was the first such mechanism to be described in 1961.

RNA Genes

A typical protein-coding gene is first copied into RNA as an intermediate in the manufacture of the final protein product. In other cases, the RNA molecules are the actual functional products, as in the synthesis of ribosomal RNA and transfer RNA. Some RNAs known as ribozymes are capable of enzymatic function, and microRNA has a regulatory role. The DNA sequences from which such RNAs are transcribed are known as non-coding RNA genes.

Some viruses store their entire genomes in the form of RNA, and contain no DNA at all. Because they use RNA to store genes, their cellular hosts may synthesize their proteins as soon as they are infected and without the delay in waiting for transcription. On the other hand, RNA retroviruses, such as HIV, require the reverse transcription of their genome from RNA into DNA before their proteins can be synthesized. RNA-mediated epigenetic inheritance has also been observed in plants and very rarely in animals.

Inheritance

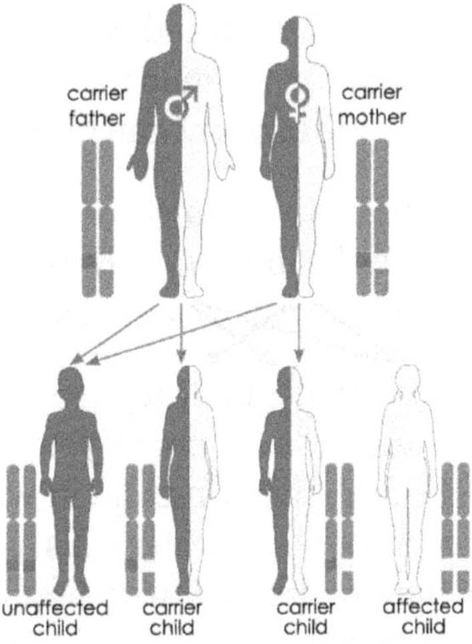

Inheritance of a gene that has two different alleles (blue and white). The gene is located on an autosomal chromosome. The blue allele is recessive to the white allele. The probability of each outcome in the children's generation is one quarter, or 25 percent.

Organisms inherit their genes from their parents. Asexual organisms simply inherit a complete copy of their parent's genome. Sexual organisms have two copies of each chromosome because they inherit one complete set from each parent.

Mendelian Inheritance

According to Mendelian inheritance, variations in an organism's phenotype (observable physical and behavioral characteristics) are due in part to variations in its genotype (particular set of genes). Each gene specifies a particular trait with different sequence of a gene (alleles) giving rise to different phenotypes. Most eukaryotic organisms (such as the pea plants Mendel worked on) have two alleles for each trait, one inherited from each parent.

Alleles at a locus may be dominant or recessive; dominant alleles give rise to their corresponding phenotypes when paired with any other allele for the same trait, whereas recessive alleles give rise to their corresponding phenotype only when paired with another copy of the same allele. For example, if the allele specifying tall stems in pea plants is dominant over the allele specifying short stems, then pea plants that inherit one tall allele from one parent and one short allele from the other parent will also have tall stems. Mendel's work demonstrated that alleles assort independently in the production of gametes, or germ cells, ensuring variation in the next generation. Although Mendelian inheritance remains a good model for many traits determined by single genes (including a number of well-known genetic disorders) it does not include the physical processes of DNA replication and cell division.

DNA Replication and Cell Division

The growth, development, and reproduction of organisms relies on cell division, or the process by which a single cell divides into two usually identical daughter cells. This requires first making a duplicate copy of every gene in the genome in a process called DNA replication. The copies are made by specialized enzymes known as DNA polymerases, which "read" one strand of the double-helical DNA, known as the template strand, and synthesize a new complementary strand. Because the DNA double helix is held together by base pairing, the sequence of one strand completely specifies the sequence of its complement; hence only one strand needs to be read by the enzyme to produce a faithful copy. The process of DNA replication is semiconservative; that is, the copy of the genome inherited by each daughter cell contains one original and one newly synthesized strand of DNA.

The rate of DNA replication in living cells was first measured as the rate of phage T4 DNA elongation in phage-infected *E. coli* and found to be impressively rapid. During the period of exponential DNA increase at 37 °C, the rate of elongation was 749 nucleotides per second.

After DNA replication is complete, the cell must physically separate the two copies of the genome and divide into two distinct membrane-bound cells. In prokaryotes (bacteria and archaea) this usually occurs via a relatively simple process called binary fission, in which each circular genome attaches to the cell membrane and is separated into the daughter cells as the membrane invaginates to split the cytoplasm into two membrane-bound portions. Binary fission is extremely fast compared to the rates of cell division in eukaryotes. Eukaryotic cell division is a more complex process known as the cell cycle; DNA replication occurs during a phase of this cycle known as S phase, whereas the process of segregating chromosomes and splitting the cytoplasm occurs during M phase.

Molecular Inheritance

The duplication and transmission of genetic material from one generation of cells to the next is the basis for molecular inheritance, and the link between the classical and molecular pictures of genes. Organisms inherit the characteristics of their parents because the cells of the offspring contain copies of the genes in their parents' cells. In asexually reproducing organisms, the offspring will be a genetic copy or clone of the parent organism. In sexually reproducing organisms, a specialized form of cell division called meiosis produces cells called gametes or germ cells that are haploid, or contain only one copy of each gene. The gametes produced by females are called eggs or ova, and those produced by males are called sperm. Two gametes fuse to form a diploid fertilized egg, a single cell that has two sets of genes, with one copy of each gene from the mother and one from the father.

During the process of meiotic cell division, an event called genetic recombination or *crossing-over* can sometimes occur, in which a length of DNA on one chromatid is swapped with a length of DNA on the corresponding homologous non-sister chromatid. This can result in reassortment of otherwise linked alleles. The Mendelian principle of independent assortment asserts that each of a parent's two genes for each trait will sort independently into gametes; which allele an organism inherits for one trait is unrelated to which allele it inherits for another trait. This is in fact only true for genes that do not reside on the same chromosome, or are located very far from one another on the same chromosome. The closer two genes lie on the same chromosome, the more closely they will be associated in gametes and the more often they will appear together; genes that are very close are essentially never separated because it is extremely unlikely that a crossover point will occur between them. This is known as genetic linkage.

Molecular Evolution

Mutation

DNA replication is for the most part extremely accurate, however errors (mutations) do occur. The error rate in eukaryotic cells can be as low as 10^{-8} per nucleotide per replication, whereas for some RNA viruses it can be as high as 10^{-3}. This means that each generation, each human genome accumulates 1–2 new mutations. Small mutations can be caused by DNA replication and the aftermath of DNA damage and include point mutations in which a single base is altered and frameshift mutations in which a single base is inserted or deleted. Either of these mutations can change the gene by missense (change a codon to encode a different amino acid) or nonsense (a premature stop codon). Larger mutations can be caused by errors in recombination to cause chromosomal abnormalities including the duplication, deletion, rearrangement or inversion of large sections of a chromosome. Additionally, DNA repair mechanisms can introduce mutational errors when repairing physical damage to the molecule. The repair, even with mutation, is more important to survival than restoring an exact copy, for example when repairing double-strand breaks.

When multiple different alleles for a gene are present in a species's population it is called polymorphic. Most different alleles are functionally equivalent, however some alleles can give rise to different phenotypic traits. A gene's most common allele is called the wild type, and rare alleles are called mutants. The genetic variation in relative frequencies of different alleles in a population is due to both natural selection and genetic drift. The wild-type allele is not necessarily the ancestor of less common alleles, nor is it necessarily fitter.

Most mutations within genes are neutral, having no effect on the organism's phenotype (silent mutations). Some mutations do not change the amino acid sequence because multiple codons encode the same amino acid (synonymous mutations). Other mutations can be neutral if they lead to amino acid sequence changes, but the protein still functions similarly with the new amino acid (e.g. conservative mutations). Many mutations, however, are deleterious or even lethal, and are removed from populations by natural selection. Genetic disorders are the result of deleterious mutations and can be due to spontaneous mutation in the affected individual, or can be inherited. Finally, a small fraction of mutations are beneficial, improving the organism's fitness and are extremely important for evolution, since their directional selection leads to adaptive evolution.

Sequence Homology

Genes with a most recent common ancestor, and thus a shared evolutionary ancestry, are known as homologs. These genes appear either from gene duplication within an organism's genome, where they are known as paralogous genes, or are the result of divergence of the genes after a speciation event, where they are known as orthologous genes, and often perform the same or similar functions in related organisms. It is often assumed that the functions of orthologous genes are more similar than those of paralogous genes, although the difference is minimal.

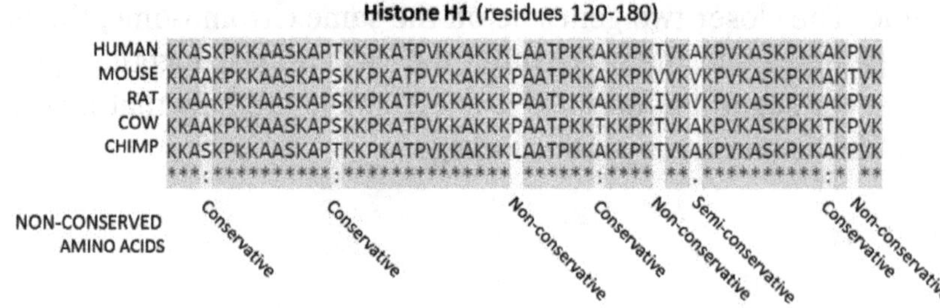

A sequence alignment, produced by ClustalO, of mammalian histone proteins

The relationship between genes can be measured by comparing the sequence alignment of their DNA. The degree of sequence similarity between homologous genes is called conserved sequence. Most changes to a gene's sequence do not affect its function and so genes accumulate mutations over time by neutral molecular evolution. Additionally, any selection on a gene will cause its sequence to diverge at a different rate. Genes under stabilizing selection are constrained and so change more slowly whereas genes under directional selection change sequence more rapidly. The sequence differences between genes can be used for phylogenetic analyses to study how those genes have evolved and how the organisms they come from are related.

Origins of New Genes

The most common source of new genes in eukaryotic lineages is gene duplication, which creates copy number variation of an existing gene in the genome. The resulting genes (paralogs) may then diverge in sequence and in function. Sets of genes formed in this way comprise a gene family. Gene duplications and losses within a family are common and represent a major source of evolutionary biodiversity. Sometimes, gene duplication may result in a nonfunctional copy of a gene, or a functional copy may be subject to mutations that result in loss of function; such nonfunctional genes are called pseudogenes.

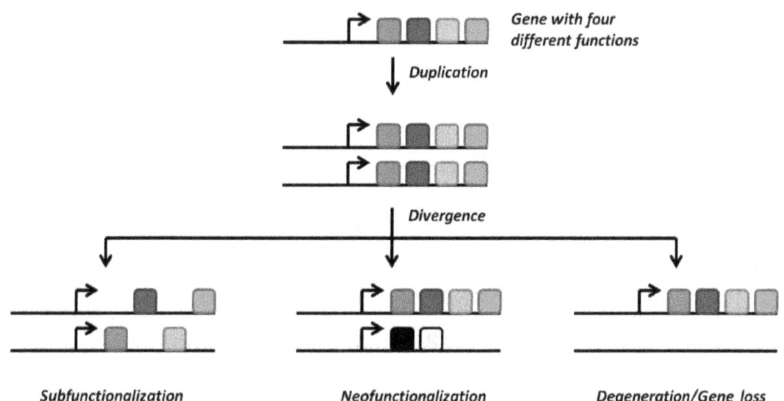

Evolutionary fate of duplicate genes

"Orphan" genes, whose sequence shows no similarity to existing genes, are less common than gene duplicates. Estimates of the number of genes with no homologs outside humans range from 18 to 60. Two primary sources of orphan protein-coding genes are gene duplication followed by extremely rapid sequence change, such that the original relationship is undetectable by sequence comparisons, and de novo conversion of a previously non-coding sequence into a protein-coding gene. De novo genes are typically shorter and simpler in structure than most eukaryotic genes, with few if any introns. Over long evolutionary time periods, de novo gene birth may be responsible for a significant fraction of taxonomically-restricted gene families.

Horizontal gene transfer refers to the transfer of genetic material through a mechanism other than reproduction. This mechanism is a common source of new genes in prokaryotes, sometimes thought to contribute more to genetic variation than gene duplication. It is a common means of spreading antibiotic resistance, virulence, and adaptive metabolic functions. Although horizontal gene transfer is rare in eukaryotes, likely examples have been identified of protist and alga genomes containing genes of bacterial origin.

Genome

The genome is the total genetic material of an organism and includes both the genes and non-coding sequences.

Number of Genes

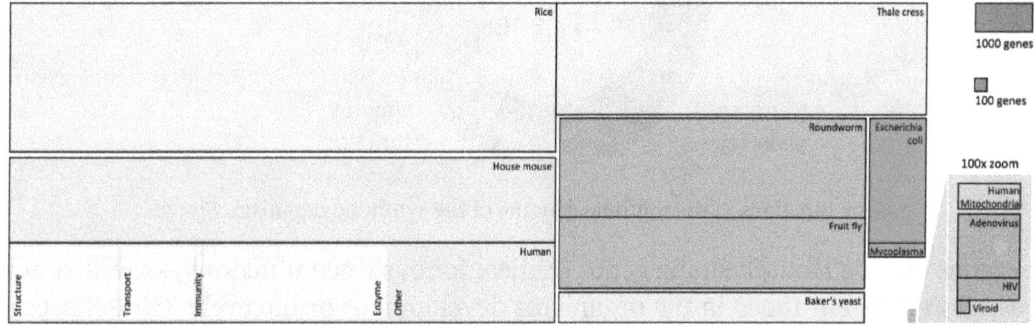

Representative genome sizes for plants (green), vertebrates (blue), invertebrates (red), fungus (yellow), bacteria (purple), and viruses (grey). An inset on the right shows the smaller genomes expanded 100-fold.

The genome size, and the number of genes it encodes varies widely between organisms. The smallest genomes occur in viruses (which can have as few as 2 protein-coding genes), and viroids (which act as a single non-coding RNA gene). Conversely, plants can have extremely large genomes, with rice containing >46,000 protein-coding genes. The total number of protein-coding genes (the Earth's proteome) is estimated to be 5 million sequences.

Although the number of base-pairs of DNA in the human genome has been known since the 1960s, the estimated number of genes has changed over time as definitions of genes, and methods of detecting them have been refined. Initial theoretical predictions of the number of human genes were as high as 2,000,000. Early experimental measures indicated there to be 50,000–100,000 *transcribed* genes (expressed sequence tags). Subsequently, the sequencing in the Human Genome Project indicated that many of these transcripts were alternative variants of the same genes, and the total number of protein-coding genes was revised down to ~20,000 with 13 genes encoded on the mitochondrial genome. Of the human genome, only 1–2% consists of protein-coding genes, with the remainder being 'noncoding' DNA such as introns, retrotransposons, and noncoding RNAs. Every multicellular organism has all its genes in each cell of its body but not every gene functions in every cell .

Essential Genes

Essential genes are the set of genes thought to be critical for an organism's survival. This definition assumes the abundant availability of all relevant nutrients and the absence of environmental stress. Only a small portion of an organism's genes are essential. In bacteria, an estimated 250–400 genes are essential for *Escherichia coli* and *Bacillus subtilis*, which is less than 10% of their genes. Half of these genes are orthologs in both organisms and are largely involved in protein synthesis. In the budding yeast *Saccharomyces cerevisiae* the number of essential genes is slightly higher, at 1000 genes (~20% of their genes). Although the number is more difficult to measure in higher eukaryotes, mice and humans are estimated to have around 2000 essential genes (~10% of their genes). The synthetic organism, *Syn 3*, has a minimal genome of 473 essential genes and quasi-essential genes (necessary for fast growth), although 149 have unknown function.

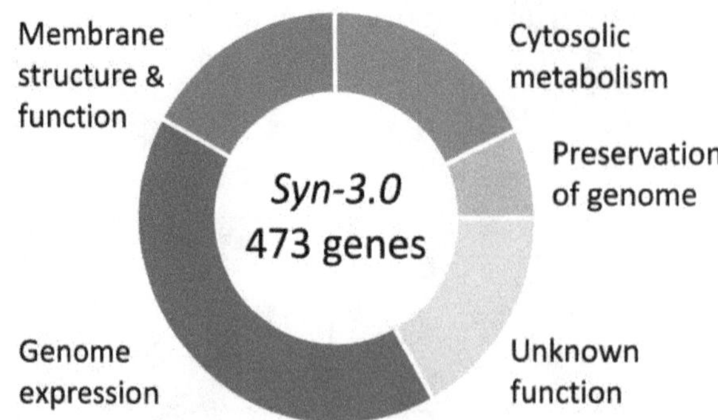

Gene functions in the minimal genome of the synthetic organism, *Syn 3*.

Essential genes include Housekeeping genes (critical for basic cell functions) as well as genes that are expressed at different times in the organisms development or life cycle. Housekeeping genes are used as experimental controls when analysing gene expression, since they are constitutively expressed at a relatively constant level.

Genetic and Genomic Nomenclature

Gene nomenclature has been established by the HUGO Gene Nomenclature Committee (HGNC) for each known human gene in the form of an approved gene name and symbol (short-form abbreviation), which can be accessed through a database maintained by HGNC. Symbols are chosen to be unique, and each gene has only one symbol (although approved symbols sometimes change). Symbols are preferably kept consistent with other members of a gene family and with homologs in other species, particularly the mouse due to its role as a common model organism.

Genetic Engineering

Genetic engineering is the modification of an organism's genome through biotechnology. Since the 1970s, a variety of techniques have been developed to specifically add, remove and edit genes in an organism. Recently developed genome engineering techniques use engineered nuclease enzymes to create targeted DNA repair in a chromosome to either disrupt or edit a gene when the break is repaired. The related term synthetic biology is sometimes used to refer to extensive genetic engineering of an organism.

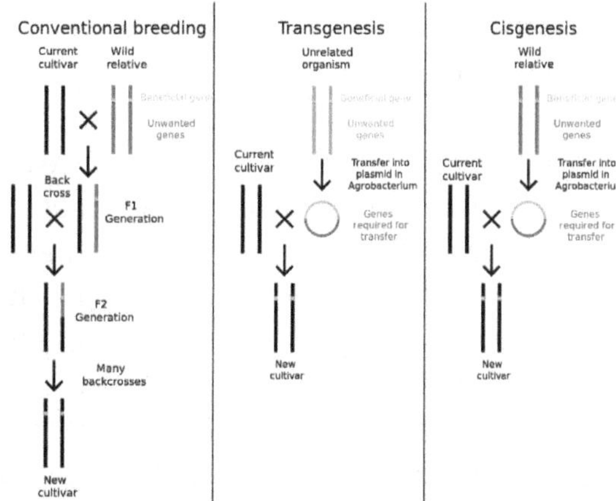

Comparison of conventional plant breeding with transgenic and cisgenic genetic modification.

Genetic engineering is now a routine research tool with model organisms. For example, genes are easily added to bacteria and lineages of knockout mice with a specific gene's function disrupted are used to investigate that gene's function. Many organisms have been genetically modified for applications in agriculture, industrial biotechnology, and medicine.

For multicellular organisms, typically the embryo is engineered which grows into the adult genetically modified organism. However, the genomes of cells in an adult organism can be edited using gene therapy techniques to treat genetic diseases.

Gene Nomenclature

Gene nomenclature is the scientific naming of genes, the units of heredity in living organisms. An international committee published recommendations for genetic symbols and nomenclature in

1957. The need to develop formal guidelines for human gene names and symbols was recognized in the 1960s and full guidelines were issued in 1979 (Edinburgh Human Genome Meeting). Several other genus-specific research communities (e.g., *Drosophila* fruit flies, *Mus* mice) have adopted nomenclature standards, as well, and have published them on the relevant model organism websites and in scientific journals, including the *Trends in Genetics* Genetic Nomenclature Guide. Scientists familiar with a particular gene family may work together to revise the nomenclature for the entire set of genes when new information becomes available. For many genes and their corresponding proteins, an assortment of alternate names is in use across the scientific literature and public biological databases, posing a challenge to effective organization and exchange of biological information. Standardization of nomenclature thus tries to achieve the benefits of vocabulary control and bibliographic control, although adherence is voluntary. The advent of the information age has brought gene ontology, which in some ways is a next step of gene nomenclature, because it aims to unify the representation of gene and gene product attributes across all species.

Gene nomenclature and protein nomenclature are not separate endeavors; they are aspects of the same whole. Any name or symbol used for a protein can potentially also be used for the gene that encodes it, and vice versa. But owing to the nature of how science has developed (with knowledge being uncovered bit by bit over decades), proteins and their corresponding genes have not always been discovered simultaneously (and not always physiologically understood when discovered), which is the largest reason why protein and gene names do not always match, or why scientists tend to favor one symbol or name for the protein and another for the gene. Another reason is that many of the mechanisms of life are the same or very similar across species, genera, orders, and phyla, so that a given protein may be produced in many kinds of organisms; and thus scientists naturally often use the same symbol and name for a given protein in one species (for example, mice) as in another species (for example, humans). Regarding the first duality (same symbol and name for gene or protein), the context usually makes the sense clear to scientific readers, and the nomenclatural systems also provide for some specificity by using italic for a symbol when the gene is meant and plain (roman) for when the protein is meant. Regarding the second duality (a given protein is endogenous in many kinds of organisms), the nomenclatural systems also provide for at least human-versus-nonhuman specificity by using different capitalization, although scientists often ignore this distinction, given that it is often biologically irrelevant.

Also owing to the nature of how scientific knowledge has unfolded, proteins and their corresponding genes often have several names and symbols that are synonymous. Some of the earlier ones may be deprecated in favor of newer ones, although such deprecation is voluntary. Some older names and symbols live on simply because they have been widely used in the scientific literature (including before the newer ones were coined) and are well established among users.

Lastly, some proteins and protein complexes are built from the products of several genes (each gene contributing a polypeptide subunit), which means that the protein or complex will not have the same name or symbol as any one gene.

Species-specific Guidelines

The HUGO Gene Nomenclature Committee is responsible for providing human gene naming guidelines and approving new, unique human gene names and symbols (short identifiers typically created by abbreviating). For some nonhuman species, model organism databases serve as central repositories of guidelines and help resources, including advice from curators and nomenclature committees. In addition to species-specific databases, approved gene names and symbols for many species can be located in the National Center for Biotechnology Information's Entrez Gene database.

Species	Guidelines	Database
Protozoa		
Dictyostelid Slime molds (*Dictyostelium discoideum*)	Nomenclature Guidelines	dictyBase
Plasmodium (*Plasmodium*)		PlasmoDB
Yeast		
Budding yeast (*Saccharomyces cerevisiae*)	SGD Gene Naming Guidelines	*Saccharomyces* Genome Database
Candida (*Candida albicans*)	*C. albicans* Gene Nomenclature Guide	*Candida* Genome Database (CGD)
Fission yeast (*Schizosaccharomyces pombe*)	Gene Name Registry	PomBase
Plants		
Maize (*Zea mays*)	A Standard For Maize Genetics Nomenclature	MaizeGDB
Thale cress (*Arabidopsis thaliana*)	Arabidopsis Nomenclature	The Arabidopsis Information Resource (TAIR).
Tree		
Flora		
Mustard (*Brassica*)	Standardized gene nomenclature for the Brassica genus (proposed)	
Animals - Invertebrates		
Fly (*Drosophila melanogaster*)	Genetic nomenclature for *Drosophila melanogaster*	FlyBase
Worm (*Caenorhabditis elegans*)	Genetic Nomenclature for *Caenorhabditis elegans* Nomenclature at a Glance Horvitz, Brenner, Hodgkin, and Herman (1979)	WormBase
Honey bee (*Apis mellifera*)		Beebase
Animals - Vertebrates		
Human (*Homo sapiens*)	Guidelines for Human Gene Nomenclature	HUGO Gene Nomenclature Committee (HGNC)
Mouse (*Mus musculus*), rat (*Rattus norvegicus*)	Rules for Nomenclature of Genes, Genetic Markers, Alleles, and Mutations in Mouse and Rat	Mouse Genome Informatics (MGI)
Anole lizard (*Anolis carolinensis*)	*Anolis* Gene Nomenclature Committee (AGNC)	AnolisGenome
Frog (*Xenopus laevis, X. tropicalis*)	Suggested Xenopus Gene Name Guidelines	Xenbase
Zebrafish (*Danio rerio*)	Zebrafish Nomenclature Guidelines	Zebrafish Model Organism Database (ZFIN)

Bacterial Genetic Nomenclature

There are generally accepted rules and conventions used for naming genes in bacteria. Standards were proposed in 1966 by Demerec et al.

General Rules

Each bacterial gene is denoted by a mnemonic of three lower case letters which indicate the pathway or process in which the gene-product is involved, followed by a capital letter signifying the actual gene. In some cases, the gene letter may be followed by an allele number. All letters and numbers are underlined or italicised. For example, *leuA* is one of the genes of the leucine biosynthetic pathway, and *leuA273* is a particular allele of this gene.

Where the actual protein coded by the gene is known then it may become part of the basis of the mnemonic, thus:

- *rpoA* encodes the α-subunit of RNA polymerase
- *rpoB* encodes the β-subunit of RNA polymerase
- *polA* encodes DNA polymerase I
- *polC* encodes DNA polymerase III
- *rpsL* encodes ribosomal protein, small S12

Some gene designations refer to a known general function:

- *dna* is involved in DNA replication

Common Mnemonics

Biosynthetic Genes

Loss of gene activity leads to a nutritional requirement (auxotrophy) not exhibited by the wildtype (prototrophy).

Amino acids:

- *ala* = alanine
- *arg* = arginine
- *asn* = asparagine

Some pathways produce metabolites that are precursors of more than one pathway. Hence, loss of one of these enzymes will lead to a requirement for more than one amino acid. For example:

- *ilv*: isoleucine and valine

Nucleotides:

- *gua* = guanine

- *pur* = purines
- *pyr* = pyrimidine
- *thy* = thymine

Vitamins:

- *bio* = biotin
- *nad* = NAD
- *pan* = pantothenic acid

Catabolic Genes

Loss of gene activity leads to loss of the ability to catabolise (use) the compound.

- *ara* = arabinose
- *gal* = galactose
- *lac* = lactose
- *mal* = maltose
- *man* = mannose
- *mel* = melibiose
- *rha* = rhamnose
- *xyl* = xylose

Drug and Bacteriophage Resistance Genes

- *amp* = ampicillin resistance
- *azi* = azide resistance
- *bla* = beta-lactam resistance
- *cat* = chloramphenicol resistance
- *kan* = kanamycin resistance
- *rif* = rifampicin resistance
- *tonA* = phage T1 resistance

Nonsense Suppressor Mutations

- *sup* = suppressor (for instance, *supF* suppresses amber mutations)

Mutant Nomenclature

If the gene in question is the wildtype a superscript '+' sign is used:

- *leuA*$^{+}$

If a gene is mutant, it is signified by a superscript '-':

- *leuA*$^{-}$

By convention, if neither is used, it is considered to be mutant.

There are additional superscripts and subscripts which provide more information about the mutation:

- ^{ts} = temperature sensitive (*leuA*^{ts})
- ^{cs} = cold sensitive (*leuA*^{cs})
- $_{am}$ = amber mutation (*leuA*$_{am}$)
- $_{um}$ = umber (opal) mutation (*leuA*$_{um}$)
- $_{oc}$ = ochre mutation (*leuA*$_{oc}$)
- ^{R} = resistant (Rif^{R})

Other modifiers:

- Δ = deletion (Δ*leuA*)
- - = fusion (*leuA-lacZ*)
- : = fusion (*leuA:lacZ*)
- :: = insertion (*leuA*::Tn*10*)
- Ω = a genetic construct introduced by a two-point crossover (Ω*leuA*)
- Δ*deleted gene*::*replacing gene* = deletion with replacement (Δ*leuA*::*nptII*(Kan^{R}) indicates that the *leuA* gene has been deleted and replaced with the gene for neomycin phosphotransferase, which confers kanamycin-resistance, as oftentimes parenthetically noted for drug-resistance markers)

Phenotype Nomenclature

When referring to the genotype (the gene) the mnemonic is italicized and not capitalised. When referring to the gene product or phenotype, the mnemonic is first-letter capitalised and not italicized (*e.g.* DnaA – the protein produced by the *dnaA* gene; LeuA^{-} – the phenotype of a *leuA* mutant; Amp^{R} – the ampicillin-resistance phenotype of the β-lactamase gene *bla*).

Bacterial Protein Name Nomenclature

Protein names are the same as the gene names, but the protein names are not italicized, and the first letter is upper-case. E.g. the name of RNA polymerase is RpoB, and this protein is encoded by *rpoB* gene.

Vertebrate Gene and Protein Symbol Conventions

Gene and protein symbol conventions ("sonic hedgehog" gene)		
Species	**Gene symbol**	**Protein symbol**
Homo sapiens	*SHH*	SHH
Mus musculus, Rattus norvegicus	*Shh*	SHH
Gallus gallus	*SHH*	SHH
Anolis carolinensis	*shh*	SHH
Xenopus laevis, X. tropicalis	*shh*	Shh
Danio rerio	*shh*	Shh

The research communities of vertebrate model organisms have adopted guidelines whereby genes in these species are given, whenever possible, the same names as their human orthologs. The use of prefixes on gene symbols to indicate species (e.g., "Z" for zebrafish) is discouraged. The recommended formatting of printed gene and protein symbols varies between species.

Symbol and Name

Vertebrate genes and proteins have names (typically strings of words) and symbols, which are short identifiers (typically 3 to 8 characters). For example, the gene cytotoxic T-lymphocyte-associated protein 4 has the HGNC symbol *CTLA4*. These symbols are usually, but not always, coined by contraction or acronymic abbreviation of the name. They are pseudo-acronyms, however, in the sense that they are complete identifiers by themselves—short names, essentially. They are synonymous with (rather than standing for) the gene/protein name (or any of its aliases), regardless of whether the initial letters "match". For example, the symbol for the gene v-akt murine thymoma viral oncogene homolog 1, which is *AKT1*, cannot be said to be an acronym for the name, and neither can any of its various synonyms, which include *AKT*, *PKB*, *PRKBA*, and *RAC*. Thus, the relationship of a gene symbol to the gene name is functionally the relationship of a nickname to a formal name (both are complete identifiers)—it is not the relationship of an acronym to its expansion. In this sense they are similar to the symbols for units of measurement in the SI system (such as km for the kilometre), in that they can be viewed as true logograms rather than just abbreviations. Sometimes the distinction is academic, but not always. Although it is not wrong to say that "VEGFA" is an acronym standing for "vascular endothelial growth factor A", just as it is not wrong that "km" is an abbreviation for "kilometre", there is more to the formality of symbols than those statements capture.

The root portion of the symbols for a gene family (such as the "SERPIN" root in *SERPIN1*, *SERPIN2*, *SERPIN3*, and so on) is called a root symbol.

Human

The HUGO Gene Nomenclature Committee is responsible for providing human gene naming guidelines and approving new, unique human gene names and symbols (short identifiers typically created by abbreviating). All human gene names and symbols can be searched at www.genenames.org, the HGNC website, and the guidelines for their formation are available there (www.genenames.org/guidelines). The guidelines for humans fit logically into the larger scope of vertebrates in general, and the HGNC's remit has recently expanded to assigning symbols to all vertebrate species without an existing nomenclature committee, to ensure that vertebrate genes are named in line with their human orthologs/paralogs. Human gene symbols generally are italicised, with all letters in uppercase (e.g., *SHH*, for sonic hedgehog). Italics are not necessary in gene catalogs. Protein designations are the same as the gene symbol, but are not italicised, with all letters in uppercase (SHH). mRNAs and cDNAs use the same formatting conventions as the gene symbol. For naming families of genes, the HGNC recommends using a "root symbol" as the root for the various gene symbols. For example, for the peroxiredoxin family, *PRDX* is the root symbol, and the family members are *PRDX1*, *PRDX2*, *PRDX3*, *PRDX4*, *PRDX5*, and *PRDX6*.

Mouse and Rat

Gene symbols generally are italicised, with only the first letter in uppercase and the remaining letters in lowercase (*Shh*). Italics are not required on web pages. Protein designations are the same as the gene symbol, but are not italicised and all are upper case (SHH).

Chicken (Gallus sp.)

Nomenclature generally follows the conventions of human nomenclature. Gene symbols generally are italicised, with all letters in uppercase (e.g., *NLGN1*, for neuroligin1). Protein designations are the same as the gene symbol, but are not italicised; all letters are in uppercase (NLGN1). mRNAs and cDNAs use the same formatting conventions as the gene symbol.

Anole Lizard (Anolis sp.)

Gene symbols are italicised and all letters are in lowercase (*shh*). Protein designations are the same as the gene symbol, are not italicised, and all letters are in uppercase (SHH).

Frog (Xenopus sp.)

Gene symbols are italicised and all letters are in lowercase (*shh*). Protein designations are the same as the gene symbol, are not italicised; the first letter is in uppercase and the remaining letters are in lowercase (Shh).

Zebrafish

Gene symbols are italicised, with all letters in lowercase (*shh*). Protein designations are the same as the gene symbol, but are not italicised; the first letter is in uppercase and the remaining letters are in lowercase (Shh).

Gene and Protein Symbol and Description in Copyediting

"Expansion" (Glossing)

A nearly universal rule in copyediting of articles for public health journals is that abbreviations and acronyms must be expanded at first use, to provide a glossing type of explanation. Typically no exceptions are permitted except for small lists of especially well known terms (such as *DNA* or *HIV*). Although readers with high subject-matter expertise do not need most of these expansions, those with intermediate or (especially) low expertise are appropriately served by them.

One complication that gene and protein symbols bring to this general rule is that they are not, accurately speaking, abbreviations or acronyms, despite the fact that many were originally coined via abbreviating or acronymic etymology. They are pseudoacronyms (as *SAT* and *KFC* also are) because they do not "stand for" any expansion. Rather, the relationship of a gene symbol to the gene name is functionally the relationship of a nickname to a formal name (both are complete identifiers)—it is not the relationship of an acronym to its expansion. In fact, many official gene symbol–gene name pairs do not even share their initial-letter sequences (although some do). Nevertheless, gene and protein symbols "look just like" abbreviations and acronyms, which presents the problem that "failing" to "expand" them (even though it is not actually a failure and there are no true expansions) creates the appearance of violating the spell-out-all-acronyms rule.

One common way of reconciling these two opposing forces is simply to exempt all gene and protein symbols from the glossing rule. This is certainly fast and easy to do, and in highly specialized journals, it is also justified because the entire target readership has high subject matter expertise. (Experts are not confused by the presence of symbols (whether known or novel) and they know where to look them up online for further details if needed.) But for journals with broader and more general target readerships, this action leaves the readers without any explanatory annotation and can leave them wondering what the apparent-abbreviation stands for and why it was not explained. Therefore a good alternative solution is simply to put either the official gene name or a suitable short description (gene alias/other designation) in parentheses after the first use of the official gene/protein symbol. This meets both the formal requirement (the presence of a gloss) and the functional requirement (helping the reader to know what the symbol refers to). The same guideline applies to shorthand names for sequence variations; AMA says, "In general medical publications, textual explanations should accompany the shorthand terms at first mention." Thus "188del11" is glossed as "an 11-bp deletion at nucleotide 188." This corollary rule (which forms an adjunct to the spell-everything-out rule) often also follows the "abbreviation-leading" style of expansion that is becoming more prevalent in recent years. Traditionally, the abbreviation always followed the fully expanded form in parentheses at first use. This is still the general rule. But for certain classes of abbreviations or acronyms (such as clinical trial acronyms [e.g., *ECOG*] or standardized polychemotherapy regimens [e.g., *CHOP*]), this pattern may be reversed, because the short form is more widely used and the expansion is merely parenthetical to the discussion at hand. The same is true of gene/protein symbols.

Synonyms and Previous Symbols and Names

The HUGO Gene Nomenclature Committee (HGNC) maintains an official symbol and name for each human gene, as well as a list of synonyms and previous symbols and names. For example, for *AFF1* (AF4/FMR2 family, member 1), previous symbols and names are *MLLT2* ("myeloid/lymphoid or mixed-lineage leukemia (trithorax (Drosophila) homolog); translocated to, 2") and *PBM1* ("pre-B-cell monocytic leukemia partner 1"), and synonyms are *AF-4* and *AF4*. Authors of journal articles often use the latest official symbol and name, but just as often they use synonyms and previous symbols and names, which are well established by earlier use in the literature. AMA style is that "authors should use the most up-to-date term" and that "in any discussion of a gene, it is recommended that the approved gene symbol be mentioned at some point, preferably in the title and abstract if relevant." Because copyeditors are not expected or allowed to rewrite the gene and protein nomenclature throughout a manuscript (except by rare express instructions on particular assignments), the middle ground in manuscripts using synonyms or older symbols is that the copyeditor will add a mention of the current official symbol at least as a parenthetical gloss at the first mention of the gene or protein, and query for confirmation.

Styling

Some basic conventions, such as (1) that animal/human homolog (ortholog) pairs differ in letter case (title case and all caps, respectively) and (2) that the symbol is italicized when referring to the gene but nonitalic when referring to the protein, are often not followed by contributors to public health journals. Many journals have the copyeditors restyle the casing and formatting to the extent feasible, although in complex genetics discussions only subject-matter experts (SMEs) can effortlessly parse them all. One example that illustrates the potential for ambiguity among non-SMEs is that some official gene names have the word "protein" within them, so the phrase "brain protein I3 (*BRI3*)" (referring to the gene) and "brain protein I3 (BRI3)" (referring to the protein) are both valid. The *AMA Manual* gives another example: both "the TH gene" and "the *TH* gene" can validly be parsed as correct ("the gene for tyrosine hydroxylase"), because the first mentions the alias (description) and the latter mentions the symbol. This seems confusing on the surface, although it is easier to understand when explained as follows: in this gene's case, as in many others, the alias (description) "happens to use the same letter string" that the symbol uses. (The matching of the letters is of course acronymic in origin and thus the phrase "happens to" implies more coincidence than is actually present; but phrasing it that way helps to make the explanation clearer.) There is no way for a non-SME to know this is the case for any particular letter string without looking up every gene from the manuscript in a database such as NCBI Gene, reviewing its symbol, name, and alias list, and doing some mental cross-referencing and double-checking (plus it helps to have biochemical knowledge). Most medical journals do not (in some cases cannot) pay for that level of fact-checking as part of their copyediting service level; therefore, it remains the author's responsibility. However, as pointed out earlier, many authors make little attempt to follow the letter case or italic guidelines; and regarding protein symbols, they often won't use the official symbol at all. For example, although the guidelines would call p53 protein "TP53" in humans or "Tp53" in rats, most authors call it "p53" in both (and even refuse to call it "TP53" if edits or queries try to), not least because of the biologic principle that many proteins are essentially or exactly the same molecules regardless of mammalian species. Regarding the gene, authors are usually willing to call it by its human-specific symbol and capitalization, *TP53*, and may even do so without being prompted by a query. But the end result of all these factors is that the published literature often does not follow the nomenclature guidelines completely.

Genome

In modern molecular biology and genetics, a genome is the genetic material of an organism. It consists of DNA (or RNA in RNA viruses). The genome includes both the genes, (the coding regions), the noncoding DNA and the genomes of the mitochondria and chloroplasts.

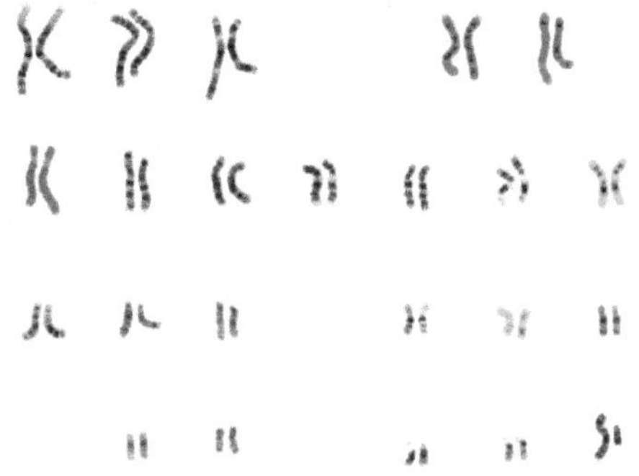

An image of the 46 chromosomes making up the diploid genome of a human male. (The mitochondrial chromosome is not shown.)

Origin of Term

The term *genome* was created in 1920 by Hans Winkler, professor of botany at the University of Hamburg, Germany. The Oxford Dictionary suggests the name is a blend of the words *gene* and *chromosome*. However, see omics for a more thorough discussion. A few related *-ome* words already existed—such as *biome*, *rhizome*, forming a vocabulary into which *genome* fits systematically.

Overview

Some organisms have multiple copies of chromosomes: diploid, triploid, tetraploid and so on. In classical genetics, in a sexually reproducing organism (typically eukarya) the gamete has half the number of chromosomes of the somatic cell and the genome is a full set of chromosomes in a diploid cell. The halving of the genetic material in gametes is accomplished by the segregation of homologous chromosomes during meiosis. In haploid organisms, including cells of bacteria, archaea, and in organelles including mitochondria and chloroplasts, or viruses, that similarly contain genes, the single or set of circular or linear chains of DNA (or RNA for some viruses), likewise constitute the genome. The term *genome* can be applied specifically to mean what is stored on a complete set of nuclear DNA (i.e., the «nuclear genome") but can also be applied to what is stored within organelles that contain their own DNA, as with the "mitochondrial genome" or the "chloroplast genome". Additionally, the genome can comprise non-chromosomal genetic elements such as viruses, plasmids, and transposable elements.

Typically, when it is said that the genome of a sexually reproducing species has been "sequenced", it refers to a determination of the sequences of one set of autosomes and one of each type of sex chromosome, which together represent both of the possible sexes. Even in species that exist in only one sex, what is described as a "genome sequence" may be a composite read from the chromosomes of various individuals. Colloquially, the phrase "genetic makeup" is sometimes used to signify the genome of a particular individual or organism. The study of the global properties of genomes of related organisms is usually referred to as genomics, which distinguishes it from genetics which generally studies the properties of single genes or groups of genes.

Both the number of base pairs and the number of genes vary widely from one species to another, and there is only a rough correlation between the two (an observation is known as the C-value paradox). At present, the highest known number of genes is around 60,000, for the protozoan causing trichomoniasis.

An analogy to the human genome stored on DNA is that of instructions stored in a book:

- The book (genome) would contain 23 chapters (chromosomes);
- Each chapter contains 48 to 250 million letters (A,C,G,T) without spaces;
- Hence, the book contains over 3.2 billion letters total;
- The book fits into a cell nucleus the size of a pinpoint;
- At least one copy of the book (all 23 chapters) is contained in most cells of our body. The only exception in humans is found in mature red blood cells which become enucleated during development and therefore lack a genome.

Sequencing and Mapping

In 1976, Walter Fiers at the University of Ghent (Belgium) was the first to establish the complete nucleotide sequence of a viral RNA-genome (Bacteriophage MS2). The next year Fred Sanger completed the first DNA-genome sequence: Phage Φ-X174, of 5386 base pairs. The first complete genome sequences among all three domains of life were released within a short period during the mid-1990s: The first bacterial genome to be sequenced was that of Haemophilus influenzae, completed by a team at The Institute for Genomic Research in 1995. A few months later, the first eukaryotic genome was completed, with sequences of the 16 chromosomes of budding yeast *Saccharomyces cerevisiae* published as the result of a European-led effort begun in the mid-1980s. The first genome sequence for an archaeon, *Methanococcus jannaschii*, was completed in 1996, again by The Institute for Genomic Research.

```
CATGACGTCGCGGACAACCCAGAATTGTCTTGAGCGATGGTAAGATCTAACCTCACTGCCGGGGGAGGCTCATAC
CTGGGGCTTTACTGATGTCATACCGTCTTGCACGGGGATAGAATGACGGTGCCCGTGTCTGCTTGCCTCGAAGCA
ATTTTCTGAAAGTTACAGACTTCGATTAAAAAGATCGGACTGCGCGTGGGCCCGGAGAGACATGCGTGGTAGTCA
TTTTTCGACGTGTCAAGGACTCAAGGGAATAGTTTGGCGGGAGCGTTACAGCTTCAATTCCCAAAGGTCGCAAGA
CGATAAAATTCAACTACTGGTTTCGGCCTAATAGGTCACGTTTTATGTGAAATAGAGGGGAACCGGCTCCCAAAT
CCCTGGGTGTTCTATGATAAGTCCTGCTTTATAACACGGGGCGGTTAGGTTAAATGACTCTTCTATCTTATGGTG
ATCCAAGCGCCCGCTAATTCTGTTCTGTTAATGTTCATACCAATACTCACATCACATTAGATCAAAGGATCCCCG
AGCCCAGTCGCAAGGGTCTGCTGCTGTTGTCGACGCCTCATGTTACTCCTGGAATCTACCTGCCCTCCCCTCACC
GGTTAAGGCGTGTGATCGACGATGCAGGTATACATCGGCTCGGACCTACAGTGGTCGATCGACTGGCTACTGGCT
TCGCGGTTCGGCGCGTAGTTGAGTGCGATAACCCAACCGGTGGCAAGTAGCAAGAAGACCTACCTGGGTCACCTT
AGACAACCTAACTAATAGTCTCTAACGGGGAATTACCTTTACCAGTCTCATGCCTCCAATATATCTGCACCGCTT
CAATGATATCGCCCACAGAAAGTAGGGTCTCAGGTATCGCATACGCCGCGCCCGGGTCCCAGCTACGCTCAGGAC
GACAGTAGAGAGCTATTGTGTAATTCAGGCTCAGCATTCATCGACCTTTCCTGTTGTGAATATTGTGCTAATGCA
TCTCGTCCGTAACGATCTGGGGGGCAAAACCGAATATCCGTATTCTCGTCCTACGGGTCCACAATGAGAAAGTCC
TGCGCGTGATCGTCAGTTAAGTTAAATTAATTCAGGCTACGGTAAACTTGTAGTGAGCTAAGAATCACGGGAATC
ACGGGTTCGCTACAGATGAACTGAATTTATACACGGACAACTCATCGCCCATTTGGGCGTGGGCACCGCAGATCA
AAAGTGGCAGATTAGGAGTGCTTGATCAGGTTAGCAGGTGGACTGTATCCAACAGCGCATCAAACTTCAATAAAT
CCAAAGCGTTGTAGTGGTCTAAGCACCCCTGAACAGTGGCGCCCATCGTTAGCGTAGTACAACCCTTCCCCCTTG
AGGTGCGACATGGGGCCAGTTAGCCTGCCCTATATCCCTTGCACACGTTCAATAAGAGGGGCTCTACAGCGCCGC
TTTTTAAATTAGGATGCCGACCCCATCATTGGTAACTGTATGTTCATAGATATTTCTTCAGGAGTAATAGCGACA
AGCTGACACGCAAGGGTCAACAATAATTTCTACTATCACCCCGCTGAACGACTGTCTTTGCAAGAACCAACTGGG
CTTAGATTCGCGTCCTAACGTAGTGAGGGCCGAGTCATATCATAGATCAGGCATGAGAAACCGACGTCGAGTCTA
CACACGAGTTGTAAACAACTTGATTGCTATACTGTAGCTACCGCAAGGATCTCCTACATCAAAGACTACTGGGCG
ATCTGGATCCGAGTCAGAAATACGAGTTAATGCAAATTTACGTAGACCGGTGAAAACACGTGCCATGGGTTGCGT
AGACCGTAGTCAGAAGTGTGGCGCGCTATTCGTACCGAACCGGTGGAGTATACAGAATTGCTCTTCTACGACGTA
AGGAGCTCGGTCCCCAATGCACGCCAAAAAAGGAATAAAGTATTCAAACTGCGCATGGTCCCTCCGCCGGTGGCA
CTATTATCCATCCGAACGTTGAACCTACTTCCTCGGCTTATGCTGTCCTCAACAGTATCGCTTATGAATCGCATG
CGGCTGTGGATCTTAACGGCCACATTCTTAATTCCGACCGATCACCGATCGCCTTTCCTCGCTGGTACAATGAGT
ACTAAGTTATCCAGATCAAGGTTTGAACGGACTCGTATGACATGTGTGACTGAACCCGGGAGGAAATGCAGAGAA
CTGTTTCAAGGCCTCTGCTTTGGTATCACTCAATATATTCAGACCAGACAAGTGGCAAAATTTCGTGCGCCTCTC
CTAGGTATTCACGCAACCGTCGTAACATGCACTAAGGATAACTAGCGCCAGGGGGGCATACTAGGTCCCGGAGCT
AAAGACTACCCTATGGATTCCTTGGAGCGGGGACAATGCAGACCGGTTACGACACAATTATCGGGATCGTCTAGA
GGTATTATTAGCAAGACAATAAAGGACATTGCACAGAGACTTATTAGAATTCAACAAACAGGATCATATCATGCG
GTGTTGGGTCGGGCAAGTCCCCGAAGCTCGGCCAAAAGATTCGCCATGGAACCGTCTGGTCCTGTTAGCGTGTAC
GCCTGCTCCTGTTCCGGGTACCATAGATAGACTGAGATTGCGTCAAAAAATTGCGGCGAAAATAGAGGGGCTCCT
TGTAGAAATACCAGACTGGGGAATTTAAGCGCTTTCCACTATCTGAGCGACTAAACATCAACAAATGCGTCTACT
CGAATCCGCAGTAGGCAATTACAACCTGGTTCAGATCACTGGTTAATCAGGGATGTCTTCATAAGATTATACTTG
CCCCGACGCGACAGCTCTTCAAGGGGCCGATTTTTGGACTTCAGATACGCTAGAATTTAAAGGGTCTCTTACACC
TGCTGCGGCCTGCAGGGACCCCTAGAACTTGCCGCCTACTTGTCTCAGTCTAATAACGCGCGAAGCCGTGGGGCA
CGTGACCTTAAGTCGCAGAGCGAGTGATGAATTTGGGACGCTAATATGGGTGAATAGAGACTTATATCATCAGGG
```

Part of DNA sequence - prototypification of complete genome of virus

The development of new technologies has made it dramatically easier and cheaper to do sequencing, and the number of complete genome sequences is growing rapidly. The US National Institutes of Health maintains one of several comprehensive databases of genomic information. Among the thousands of completed genome sequencing projects include those for rice, a mouse, the plant *Arabidopsis thaliana*, the puffer fish, and the bacteria E. coli. In December 2013, scientists first sequenced the entire *genome* of a Neanderthal, an extinct species of humans. The genome was extracted from the toe bone of a 130,000-year-old Neanderthal found in a Siberian cave.

New sequencing technologies, such as massive parallel sequencing have also opened up the prospect of personal genome sequencing as a diagnostic tool, as pioneered by Manteia Predictive Medicine. A major step toward that goal was the completion in 2007 of the full genome of James D. Watson, one of the co-discoverers of the structure of DNA.

Whereas a genome sequence lists the order of every DNA base in a genome, a genome map identifies the landmarks. A genome map is less detailed than a genome sequence and aids in navigating around the genome. The Human Genome Project was organized to map and to sequence the human genome. A fundamental step in the project was the release of a detailed genomic map by Jean Weissenbach and his team at the Genoscope in Paris.

Reference genome sequences and maps continue to be updated, removing errors and clarifying regions of high allelic complexity. The decreasing cost of genomic mapping has permitted genealogical sites to offer it as a service, to the extent that one may submit one's genome to crowd sourced scientific endeavours such as DNA.land at the New York Genome Center, an example both of the economies of scale and of citizen science.

Genome Compositions

Genome composition is used to describe the make up of contents of a haploid genome, which should include genome size, proportions of non-repetitive DNA and repetitive DNA in details. By comparing the genome compositions between genomes, scientists can better understand the evolutionary history of a given genome.

When talking about genome composition, one should distinguish between prokaryotes and eukaryotes as there are significant differences with contents structure. In prokaryotes, most of the genome (85–90%) is non-repetitive DNA, which means coding DNA mainly forms it, while non-coding regions only take a small part. On the contrary, eukaryotes have the feature of exon-intron organization of protein coding genes; the variation of repetitive DNA content in eukaryotes is also extremely high. In mammals and plants, the major part of the genome is composed of repetitive DNA.

Most biological entities that are more complex than a virus sometimes or always carry additional genetic material besides that which resides in their chromosomes. In some contexts, such as sequencing the genome of a pathogenic microbe, "genome" is meant to include information stored on this auxiliary material, which is carried in plasmids. In such circumstances then, "genome" describes all of the genes and information on non-coding DNA that have the potential to be present.

In eukaryotes such as plants, protozoa and animals, however, "genome" carries the typical connotation of only information on chromosomal DNA. So although these organisms contain chloroplasts or mitochondria that have their own DNA, the genetic information contained in DNA within these organelles is not considered part of the genome. In fact, mitochondria are sometimes said to have their own genome often referred to as the "mitochondrial genome". The DNA found within the chloroplast may be referred to as the "plastome".

Genome Size

Genome size is the total number of DNA base pairs in one copy of a haploid genome. The nuclear genome comprises approximately 3.2 billion nucleotides of DNA, divided into 24 linear molecules, the shortest 50 000 000 nucleotides in length and the longest 260 000 000 nucleotides, each contained in a different chromosome. The genome size is positively correlated with the morphological complexity among prokaryotes and lower eukaryotes; however, after mollusks and all the other higher eukaryotes above, this correlation is no longer effective. This phenomenon also indicates the mighty influence coming from repetitive DNA act on the genomes.

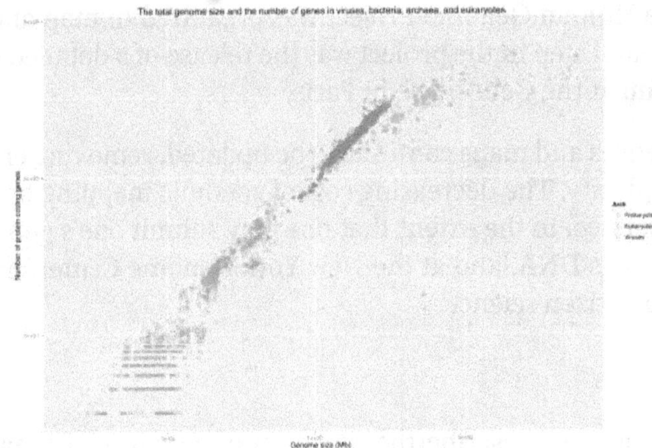

Log-log plot of the total number of annotated proteins in genomes submitted to GenBank as a function of genome size.

Since genomes are very complex, one research strategy is to reduce the number of genes in a genome to the bare minimum and still have the organism in question survive. There is experimental work being done on minimal genomes for single cell organisms as well as minimal genomes for multi-cellular organisms. The work is both *in vivo* and *in silico*.

Here is a table of some significant or representative genomes.

Organism type	Organism	Genome size (base pairs)		Approx. no. of genes	Note
Virus	Porcine circovirus type 1	1,759	1.8kb		Smallest viruses replicating autonomously in eukaryotic cells.
Virus	Bacteriophage MS2	3,569	3.5kb		First sequenced RNA-genome
Virus	SV40	5,224	5.2kb		
Virus	Phage Φ-X174	5,386	5.4kb		First sequenced DNA-genome
Virus	HIV	9,749	9.7kb		
Virus	Phage λ	48,502	48kb		Often used as a vector for the cloning of recombinant DNA.

Virus	Megavirus	1,259,197	1.3Mb		Until 2013 the largest known viral genome.
Virus	*Pandoravirus salinus*	2,470,000	2.47Mb		Largest known viral genome.
Bacterium	*Nasuia deltocephalinicola* (strain NAS-ALF)	112,091	112kb		Smallest non-viral genome.
Bacterium	*Carsonella ruddii*	159,662	160kb		
Bacterium	*Buchnera aphidicola*	600,000	600kb		
Bacterium	*Wigglesworthia glossinidia*	700,000	700Kb		
Bacterium	*Haemophilus influenzae*	1,830,000	1.8Mb		First genome of a living organism sequenced, July 1995
Bacterium	*Escherichia coli*	4,600,000	4.6Mb	4288	
Bacterium	*Solibacter usitatus* (strain Ellin 6076)	9,970,000	10Mb		
Bacterium – cyanobacterium	*Prochlorococcus* spp. (1.7 Mb)	1,700,000	1.7Mb	1884	Smallest known cyanobacterium genome
Bacterium – cyanobacterium	*Nostoc punctiforme*	9,000,000	9Mb	7432	7432 "open reading frames"
Amoeboid	*Polychaos dubium* ("Amoeba" dubia)	670,000,000,000	670Gb		Largest known genome. (Disputed)
Plant	*Genlisea tuberosa*	61,000,000	61Mb		Smallest recorded flowering plant genome, 2014.
Plant	*Arabidopsis thaliana*	157,000,000	157Mb	25498	First plant genome sequenced, December 2000.
Plant	*Populus trichocarpa*	480,000,000	480Mb	73013	First tree genome sequenced, September 2006
Plant	*Fritillaria assyrica*	130,000,000,000	130Gb		
Plant	*Paris japonica* (Japanese-native, pale-petal)	150,000,000,000	150Gb		Largest plant genome known
Plant – moss	*Physcomitrella patens*	480,000,000	480Mb		First genome of a bryophyte sequenced, January 2008.
Fungus – yeast	*Saccharomyces cerevisiae*	12,100,000	12.1Mb	6294	First eukaryotic genome sequenced, 1996
Fungus	*Aspergillus nidulans*	30,000,000	30Mb	9541	
Nematode	*Pratylenchus coffeae*	20,000,000	20Mb		Smallest animal genome known
Nematode	*Caenorhabditis elegans*	100,300,000	100Mb	19000	First multicellular animal genome sequenced, December 1998
Insect	*Drosophila melanogaster* (fruit fly)	175,000,000	175Mb	13600	Size variation based on strain (175-180Mb; standard *y w* strain is 175Mb)

Insect	*Apis mellifera* (honey bee)	236,000,000	236Mb	10157	)
Insect	*Bombyx mori* (silk moth)	432,000,000	432Mb	14623	14,623 predicted genes
Insect	*Solenopsis invicta* (fire ant)	480,000,000	480Mb	16569	
Mammal	*Mus musculus*	2,700,000,000	2.7Gb	20210	
Mammal	*Homo sapiens*	3,289,000,000	3.3Gb	20000	*Homo sapiens* estimated genome size 3.2 billion bp Initial sequencing and analysis of the human genome
Mammal	*Bonobo*	3,286,640,000	3.3Gb	20000	*Pan paniscus* estimated genome size 3.29 billion bp
Fish	*Tetraodon nigroviridis* (type of puffer fish)	385,000,000	390Mb		Smallest vertebrate genome known estimated to be 340 Mb – 385 Mb.
Fish	*Protopterus aethiopicus* (marbled lungfish)	130,000,000,000	130Gb		Largest vertebrate genome known

Proportion of Non-repetitive DNA

The proportion of non-repetitive DNA is calculated by using the length of non-repetitive DNA divided by genome size. Protein-coding genes and RNA-coding genes are generally non-repetitive DNA. A bigger genome does not mean more genes, and the proportion of non-repetitive DNA decreases along with increasing genome size in higher eukaryotes.

It had been found that the proportion of non-repetitive DNA can vary a lot between species. Some *E. coli* as prokaryotes only have non-repetitive DNA, lower eukaryotes such as *C. elegans* and fruit fly, still possess more non-repetitive DNA than repetitive DNA. Higher eukaryotes tend to have more repetitive DNA than non-repetitive ones. In some plants and amphibians, the proportion of non-repetitive DNA is no more than 20%, becoming a minority component.

Proportion of Repetitive DNA

The proportion of repetitive DNA is calculated by using length of repetitive DNA divide by genome size. There are two categories of repetitive DNA in genome: tandem repeats and interspersed repeats.

Tandem Repeats

Tandem repeats are usually caused by slippage during replication, unequal crossing-over and gene conversion, satellite DNA and microsatellites are forms of tandem repeats in the genome. Although tandem repeats count for a significant proportion in genome, the largest proportion in mammalian is the other type, interspersed repeats.

Interspersed Repeats

Interspersed repeats mainly come from transposable elements (TEs), but they also include some protein coding gene families and pseudogenes. Transposable elements are able to integrate into the genome at another site within the cell. It is believed that TEs are an important driving force on genome evolution of higher eukaryotes. TEs can be classified into two categories, Class 1 (retrotransposons) and Class 2 (DNA transposons).

Retrotransposons

Retrotransposons can be transcribed into RNA, which are then duplicated at another site into the genome. Retrotransposons can be divided into Long terminal repeats (LTRs) and Non-Long Terminal Repeats (Non-LTR).

Long Terminal Repeats (LTRs)

> similar to retroviruses, which have both gag and pol genes to make cDNA from RNA and proteins to insert into genome, but LTRs can only act within the cell as they lack the env gene in retroviruses. It has been reported that LTRs consist of the largest fraction in most plant genome and might account for the huge variation in genome size.

Non-Long Terminal Repeats (Non-LTRs)

> can be divided into long interspersed elements (LINEs), short interspersed elements (SINEs) and Penelope-like elements. In *Dictyostelium discoideum*, there is another DIRS-like elements belong to Non-LTRs. Non-LTRs are widely spread in eukaryotic genomes.

Long interspersed elements (LINEs)

> are able to encode two Open Reading Frames (ORFs) to generate transcriptase and endonuclease, which are essential in retrotransposition. The human genome has around 500,000 LINEs, taking around 17% of the genome.

Short interspersed elements (SINEs)

> are usually less than 500 base pairs and need to co-opt with the LINEs machinery to function as nonautonomous retrotransposons. The Alu element is the most common SINEs found in primates, it has a length of about 350 base pairs and takes about 11% of the human genome with around 1,500,000 copies.

DNA Transposons

DNA transposons generally move by "cut and paste" in the genome, but duplication has also been observed. Class 2 TEs do not use RNA as intermediate and are popular in bacteria, in metazoan it has also been found.

Genome Evolution

Genomes are more than the sum of an organism's genes and have traits that may be measured and studied without reference to the details of any particular genes and their products. Researchers compare traits such as *chromosome number* (karyotype), genome size, gene order, codon usage bias, and GC-content to determine what mechanisms could have produced the great variety of genomes that exist today.

Duplications play a major role in shaping the genome. Duplication may range from extension of short tandem repeats, to duplication of a cluster of genes, and all the way to duplication of entire chromosomes or even entire genomes. Such duplications are probably fundamental to the creation of genetic novelty.

Horizontal gene transfer is invoked to explain how there is often an extreme similarity between small portions of the genomes of two organisms that are otherwise very distantly related. Horizontal gene transfer seems to be common among many microbes. Also, eukaryotic cells seem to have experienced a transfer of some genetic material from their chloroplast and mitochondrial genomes to their nuclear chromosomes.

Gene Expression

Gene expression is the process by which information from a gene is used in the synthesis of a functional gene product. These products are often proteins, but in non-protein coding genes such as transfer RNA (tRNA) or small nuclear RNA (snRNA) genes, the product is a functional RNA. The process of gene expression is used by all known life—eukaryotes (including multicellular organisms), prokaryotes (bacteria and archaea), and utilized by viruses—to generate the macromolecular machinery for life.

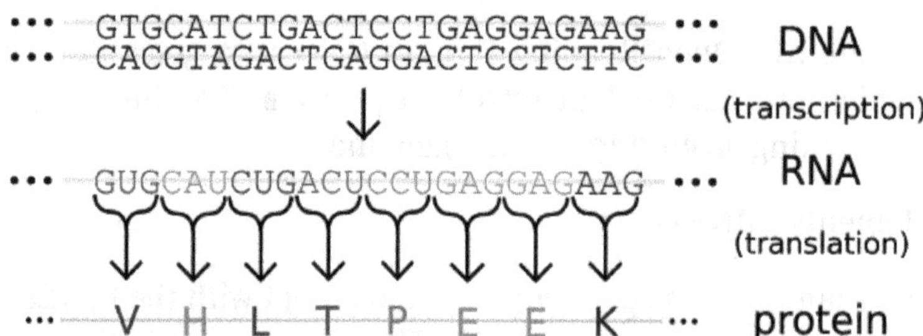

Genes are expressed by being transcribed into RNA, and this transcript may then be translated into protein.

Several steps in the gene expression process may be modulated, including the transcription, RNA splicing, translation, and post-translational modification of a protein. Gene regulation gives the cell control over structure and function, and is the basis for cellular differentiation, morphogenesis and the versatility and adaptability of any organism. Gene regulation may also serve as a substrate for evolutionary change, since control of the timing, location, and amount of gene expression can have a profound effect on the functions (actions) of the gene in a cell or in a multicellular organism.

In genetics, gene expression is the most fundamental level at which the genotype gives rise to the phenotype, i.e. observable trait. The genetic code stored in DNA is "interpreted" by gene expression, and the properties of the expression give rise to the organism's phenotype. Such phenotypes are often expressed by the synthesis of proteins that control the organism's shape, or that act as enzymes catalysing specific metabolic pathways characterising the organism. Regulation of gene expression is thus critical to an organism's development.

Mechanism

Transcription

A gene is a stretch of DNA that encodes information. Genomic DNA consists of two antiparallel and reverse complementary strands, each having 5' and 3' ends. With respect to a gene, the two strands may be labeled the "template strand," which serves as a blueprint for the production of an RNA transcript, and the "coding strand," which includes the DNA version of the transcript sequence. (Perhaps surprisingly, the "coding strand" is not physically involved in the coding process because it is the "template strand" that is read during transcription.)

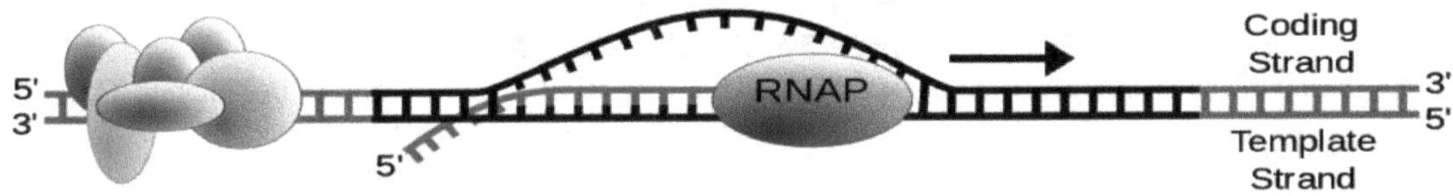

The process of transcription is carried out by RNA polymerase (RNAP), which uses DNA (black) as a template and produces RNA (blue).

The production of the RNA copy of the DNA is called transcription, and is performed in the nucleus by RNA polymerase, which adds one RNA nucleotide at a time to a growing RNA strand as per the complementarity law of the bases. This RNA is complementary to the template 3' → 5' DNA strand, which is itself complementary to the coding 5' → 3' DNA strand. Therefore, the resulting 5' → 3' RNA strand is identical to the coding DNA strand with the exception that thymines (T) are replaced with uracils (U) in the RNA. A coding DNA strand reading "ATG" is indirectly transcribed through the non-coding strand as "AUG" in RNA.

In prokaryotes, transcription is carried out by a single type of RNA polymerase, which needs a DNA sequence called a Pribnow box as well as a sigma factor (σ factor) to start transcription. In eukaryotes, transcription is performed by three types of RNA polymerases, each of which needs a special DNA sequence called the promoter and a set of DNA-binding proteins—transcription factors—to initiate the process. RNA polymerase I is responsible for transcription of ribosomal RNA (rRNA) genes. RNA polymerase II (Pol II) transcribes all protein-coding genes but also some non-coding RNAs (e.g., snRNAs, snoRNAs or long non-coding RNAs). Pol II includes a C-terminal domain (CTD) that is rich in serine residues. When these residues are phosphorylated, the CTD binds to various protein factors that promote transcript maturation and modification. RNA polymerase III transcribes 5S rRNA, transfer RNA (tRNA) genes, and some small non-coding RNAs (e.g., 7SK). Transcription ends when the polymerase encounters a sequence called the terminator.

RNA Processing

While transcription of prokaryotic protein-coding genes creates messenger RNA (mRNA) that is ready for translation into protein, transcription of eukaryotic genes leaves a primary transcript of RNA (pre-mRNA), which first has to undergo a series of modifications to become a mature mRNA.

These include 5' *capping*, which is set of enzymatic reactions that add 7-methylguanosine (m^7G) to the 5' end of pre-mRNA and thus protect the RNA from degradation by exonucleases. The m^7G cap is then bound by cap binding complex heterodimer (CBC20/CBC80), which aids in mRNA export to cytoplasm and also protect the RNA from decapping.

Another modification is 3' *cleavage and polyadenylation.* They occur if polyadenylation signal sequence (5'- AAUAAA-3') is present in pre-mRNA, which is usually between protein-coding sequence and terminator. The pre-mRNA is first cleaved and then a series of ~200 adenines (A) are added to form poly(A) tail, which protects the RNA from degradation. Poly(A) tail is bound by multiple poly(A)-binding proteins (PABP) necessary for mRNA export and translation re-initiation.

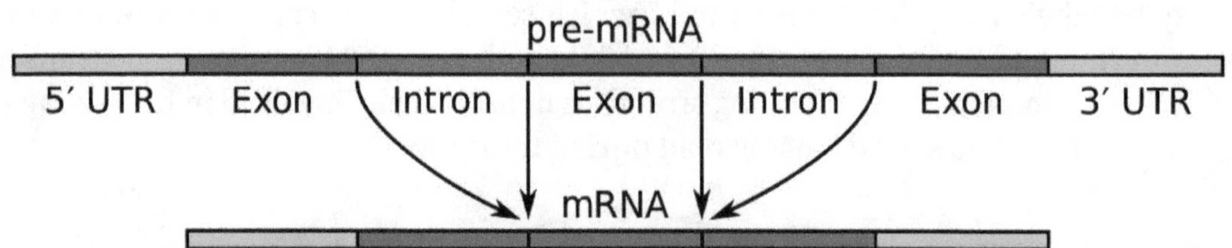

Simple illustration of exons and introns in pre-mRNA and the formation of mature mRNA by splicing. The UTRs are non-coding parts of exons at the ends of the mRNA.

A very important modification of eukaryotic pre-mRNA is *RNA splicing.* The majority of eukaryotic pre-mRNAs consist of alternating segments called exons and introns. During the process of splicing, an RNA-protein catalytical complex known as spliceosome catalyzes two transesterification reactions, which remove an intron and release it in form of lariat structure, and then splice neighbouring exons together. In certain cases, some introns or exons can be either removed or retained in mature mRNA. This so-called alternative splicing creates series of different transcripts originating from a single gene. Because these transcripts can be potentially translated into different proteins, splicing extends the complexity of eukaryotic gene expression.

Extensive RNA processing may be an evolutionary advantage made possible by the nucleus of eukaryotes. In prokaryotes, transcription and translation happen together, whilst in eukaryotes, the nuclear membrane separates the two processes, giving time for RNA processing to occur.

Non-coding RNA Maturation

In most organisms non-coding genes (ncRNA) are transcribed as precursors that undergo further processing. In the case of ribosomal RNAs (rRNA), they are often transcribed as a pre-rRNA that contains one or more rRNAs. The pre-rRNA is cleaved and modified (2′-O-methylation and pseudouridine formation) at specific sites by approximately 150 different small nucleolus-restricted RNA species, called snoRNAs. SnoRNAs associate with proteins, forming snoRNPs. While snoRNA part basepair with the target RNA and thus position the modification at a precise site, the protein part performs the catalytical reaction. In eukaryotes, in particular a snoRNP called RNase, MRP cleaves the 45S pre-rRNA into the 28S, 5.8S, and 18S rRNAs. The rRNA and RNA processing factors form large aggregates called the nucleolus.

In the case of transfer RNA (tRNA), for example, the 5' sequence is removed by RNase P, whereas the 3' end is removed by the tRNase Z enzyme and the non-templated 3' CCA tail is added by a nucleotidyl transferase. In the case of micro RNA (miRNA), miRNAs are first transcribed as primary transcripts or pri-miRNA with a cap and poly-A tail and processed to short, 70-nucleotide stem-loop structures known as pre-miRNA in the cell nucleus by the enzymes Drosha and Pasha. After being exported, it is then processed to mature miRNAs in the cytoplasm by interaction with the endonuclease Dicer, which also initiates the formation of the RNA-induced silencing complex (RISC), composed of the Argonaute protein.

Even snRNAs and snoRNAs themselves undergo series of modification before they become part of functional RNP complex. This is done either in the nucleoplasm or in the specialized compartments called Cajal bodies. Their bases are methylated or pseudouridinilated by a group of small Cajal body-specific RNAs (scaRNAs), which are structurally similar to snoRNAs.

RNA Export

In eukaryotes most mature RNA must be exported to the cytoplasm from the nucleus. While some RNAs function in the nucleus, many RNAs are transported through the nuclear pores and into the cytosol. Notably this includes all RNA types involved in protein synthesis. In some cases RNAs are additionally transported to a specific part of the cytoplasm, such as a synapse; they are then towed by motor proteins that bind through linker proteins to specific sequences (called "zipcodes") on the RNA.

Translation

For some RNA (non-coding RNA) the mature RNA is the final gene product. In the case of messenger RNA (mRNA) the RNA is an information carrier coding for the synthesis of one or more proteins. mRNA carrying a single protein sequence (common in eukaryotes) is monocistronic whilst mRNA carrying multiple protein sequences (common in prokaryotes) is known as polycistronic.

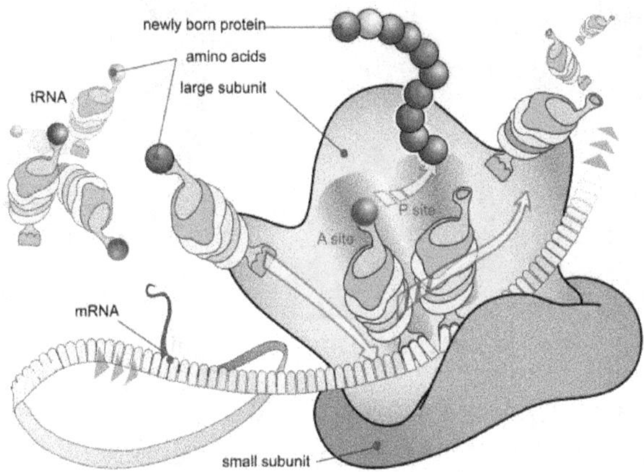

During the translation, tRNA charged with amino acid enters the ribosome and aligns with the correct mRNA triplet. Ribosome then adds amino acid to growing protein chain.

Every mRNA consists of three parts: a 5' untranslated region (5'UTR), a protein-coding region or open reading frame (ORF), and a 3' untranslated region (3'UTR). The coding region carries information for protein synthesis encoded by the genetic code to form triplets. Each triplet of nucleotides of the coding region is called a codon and corresponds to a binding site complementary to an anticodon triplet in transfer RNA. Transfer RNAs with the same anticodon sequence always carry an identical type of amino acid. Amino acids are then chained together by the ribosome according to the order of triplets in the coding region. The ribosome helps transfer RNA to bind to messenger RNA and takes the amino acid from each transfer RNA and makes a structure-less protein out of it. Each mRNA molecule is translated into many protein molecules, on average ~2800 in mammals.

In prokaryotes translation generally occurs at the point of transcription (co-transcriptionally), often using a messenger RNA that is still in the process of being created. In eukaryotes translation can occur in a variety of regions of the cell depending on where the protein being written is supposed to be. Major locations are the cytoplasm for soluble cytoplasmic proteins and the membrane of the endoplasmic reticulum for proteins that are for export from the cell or insertion into a cell membrane. Proteins that are supposed to be expressed at the endoplasmic reticulum are recognised part-way through the translation process. This is governed by the signal recognition particle—a protein that binds to the ribosome and directs it to the endoplasmic reticulum when it finds a signal peptide on the growing (nascent) amino acid chain. Translation is the communication of the meaning of a source-language text by means of an equivalent target-language text

Folding

The polypeptide folds into its characteristic and functional three-dimensional structure from a random coil. Each protein exists as an unfolded polypeptide or random coil when translated from a sequence of mRNA into a linear chain of amino acids. This polypeptide lacks any developed three-dimensional structure (the left hand side of the neighboring figure). Amino acids interact with each other to produce a well-defined three-dimensional structure, the folded protein (the right hand side of the figure) known as the native state. The resulting three-dimensional structure is determined by the amino acid sequence (Anfinsen's dogma).

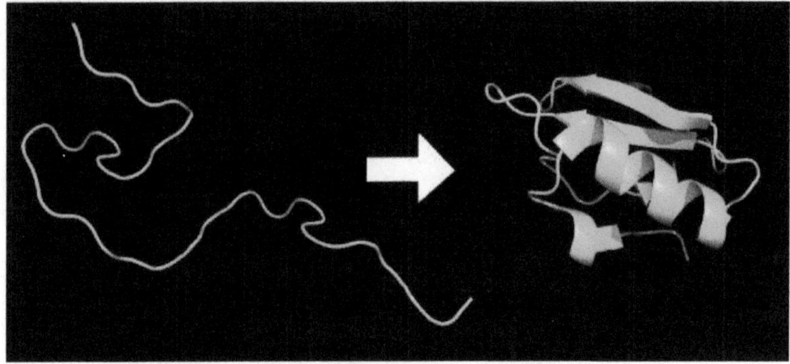

Protein before (left) and after (right) folding.

The correct three-dimensional structure is essential to function, although some parts of functional proteins may remain unfolded Failure to fold into the intended shape usually produces inactive proteins with different properties including toxic prions. Several neurodegenerative and other diseases are believed to result from the accumulation of *misfolded* proteins. Many allergies are caused by the folding of the proteins, for the immune system does not produce antibodies for certain protein structures.

Enzymes called chaperones assist the newly formed protein to attain (fold into) the 3-dimensional structure it needs to function. Similarly, RNA chaperones help RNAs attain their functional shapes. Assisting protein folding is one of the main roles of the endoplasmic reticulum in eukaryotes.

Translocation

Secretory proteins of eukaryotes or prokaryotes must be translocated to enter the secretory pathway. Newly synthesized proteins are directed to the eukaryotic Sec61 or prokaryotic SecYEG translocation channel by signal peptides. The efficiency of protein secretion in eukaryotes is very dependent on the signal peptide which has been used.

Protein Transport

Many proteins are destined for other parts of the cell than the cytosol and a wide range of signalling sequences or (signal peptides) are used to direct proteins to where they are supposed to be. In prokaryotes this is normally a simple process due to limited compartmentalisation of the cell. However, in eukaryotes there is a great variety of different targeting processes to ensure the protein arrives at the correct organelle.

Not all proteins remain within the cell and many are exported, for example, digestive enzymes, hormones and extracellular matrix proteins. In eukaryotes the export pathway is well developed and the main mechanism for the export of these proteins is translocation to the endoplasmic reticulum, followed by transport via the Golgi apparatus.

Regulation of Gene Expression

The patchy colours of a tortoiseshell cat are the result of different levels of expression of pigmentation genes in different areas of the skin.

Regulation of gene expression refers to the control of the amount and timing of appearance of the functional product of a gene. Control of expression is vital to allow a cell to produce the gene products it needs when it needs them; in turn, this gives cells the flexibility to adapt to a variable environment, external signals, damage to the cell, etc. Some simple examples of where gene expression is important are:

- Control of insulin expression so it gives a signal for blood glucose regulation.
- X chromosome inactivation in female mammals to prevent an “overdose” of the genes it contains.
- Cyclin expression levels control progression through the eukaryotic cell cycle.

More generally, gene regulation gives the cell control over all structure and function, and is the basis for cellular differentiation, morphogenesis and the versatility and adaptability of any organism.

Any step of gene expression may be modulated, from the DNA-RNA transcription step to post-translational modification of a protein. The stability of the final gene product, whether it is

RNA or protein, also contributes to the expression level of the gene—an unstable product results in a low expression level. In general gene expression is regulated through changes in the number and type of interactions between molecules that collectively influence transcription of DNA and translation of RNA.

Numerous terms are used to describe types of genes depending on how they are regulated; these include:

- A constitutive gene is a gene that is transcribed continually as opposed to a facultative gene, which is only transcribed when needed.
- A *housekeeping gene* is typically a constitutive gene that is transcribed at a relatively constant level. The housekeeping gene's products are typically needed for maintenance of the cell. It is generally assumed that their expression is unaffected by experimental conditions. Examples include actin, GAPDH and ubiquitin.
- A facultative gene is a gene only transcribed when needed as opposed to a constitutive gene.
- An inducible gene is a gene whose expression is either responsive to environmental change or dependent on the position in the cell cycle.

Transcriptional Regulation

Regulation of transcription can be broken down into three main routes of influence; genetic (direct interaction of a control factor with the gene), modulation interaction of a control factor with the transcription machinery and epigenetic (non-sequence changes in DNA structure that influence transcription).

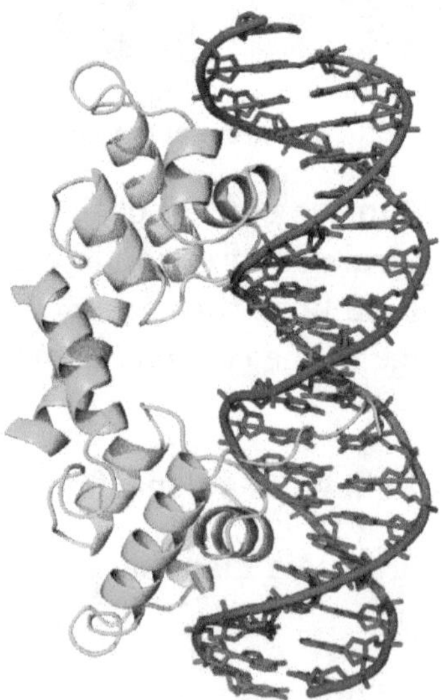

The lambda repressor transcription factor (green) binds as a dimer to major groove of DNA target (red and blue) and disables initiation of transcription. From PDB: 1LMB.

Direct interaction with DNA is the simplest and the most direct method by which a protein changes transcription levels. Genes often have several protein binding sites around the coding region with the specific function of regulating transcription. There are many classes of regulatory DNA binding sites known as enhancers, insulators and silencers. The mechanisms for regulating transcription are very varied, from blocking key binding sites on the DNA for RNA polymerase to acting as an activator and promoting transcription by assisting RNA polymerase binding.

The activity of transcription factors is further modulated by intracellular signals causing protein post-translational modification including phosphorylated, acetylated, or glycosylated. These changes influence a transcription factor's ability to bind, directly or indirectly, to promoter DNA, to recruit RNA polymerase, or to favor elongation of a newly synthesized RNA molecule.

The nuclear membrane in eukaryotes allows further regulation of transcription factors by the duration of their presence in the nucleus, which is regulated by reversible changes in their structure and by binding of other proteins. Environmental stimuli or endocrine signals may cause modification of regulatory proteins eliciting cascades of intracellular signals, which result in regulation of gene expression.

More recently it has become apparent that there is a significant influence of non-DNA-sequence specific effects on translation. These effects are referred to as epigenetic and involve the higher order structure of DNA, non-sequence specific DNA binding proteins and chemical modification of DNA. In general epigenetic effects alter the accessibility of DNA to proteins and so modulate transcription.

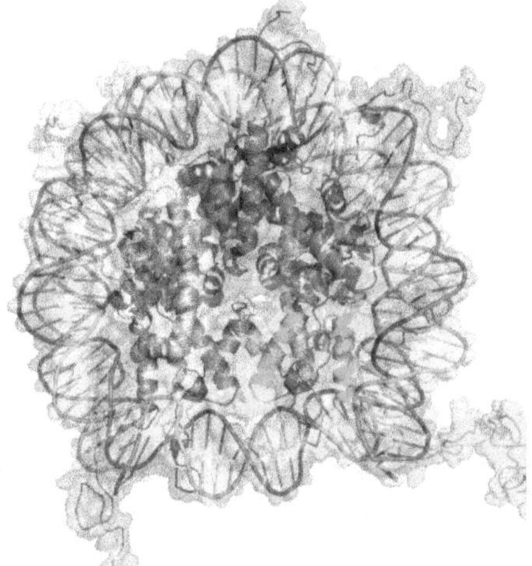

In eukaryotes, DNA is organized in form of nucleosomes. Note how the DNA (blue and green) is tightly wrapped around the protein core made of histone octamer (ribbon coils), restricting access to the DNA. From PDB: 1KX5.

DNA methylation is a widespread mechanism for epigenetic influence on gene expression and is seen in bacteria and eukaryotes and has roles in heritable transcription silencing and transcription regulation. In eukaryotes the structure of chromatin, controlled by the histone code, regulates access to DNA with significant impacts on the expression of genes in euchromatin and heterochromatin areas.

Transcriptional Regulation in Cancer

The majority of gene promoters contain a CpG island with numerous CpG sites. When many of a gene's promoter CpG sites are methylated the gene becomes silenced. Colorectal cancers typically have 3 to 6 driver mutations and 33 to 66 hitchhiker or passenger mutations. However, transcriptional silencing may be of more importance than mutation in causing progression to cancer. For example, in colorectal cancers about 600 to 800 genes are transcriptionally silenced by CpG island methylation (see regulation of transcription in cancer). Transcriptional repression in cancer can also occur by other epigenetic mechanisms, such as altered expression of microRNAs. In breast cancer, transcriptional repression of BRCA1 may occur more frequently by over-expressed microRNA-182 than by hypermethylation of the BRCA1 promoter

Post-transcriptional Regulation

In eukaryotes, where export of RNA is required before translation is possible, nuclear export is thought to provide additional control over gene expression. All transport in and out of the nucleus is via the nuclear pore and transport is controlled by a wide range of importin and exportin proteins.

Expression of a gene coding for a protein is only possible if the messenger RNA carrying the code survives long enough to be translated. In a typical cell, an RNA molecule is only stable if specifically protected from degradation. RNA degradation has particular importance in regulation of expression in eukaryotic cells where mRNA has to travel significant distances before being translated. In eukaryotes, RNA is stabilised by certain post-transcriptional modifications, particularly the 5' cap and poly-adenylated tail.

Intentional degradation of mRNA is used not just as a defence mechanism from foreign RNA (normally from viruses) but also as a route of mRNA *destabilisation*. If an mRNA molecule has a complementary sequence to a small interfering RNA then it is targeted for destruction via the RNA interference pathway.

Three Prime Untranslated Regions and microRNAs

Three prime untranslated regions (3'UTRs) of messenger RNAs (mRNAs) often contain regulatory sequences that post-transcriptionally influence gene expression. Such 3'-UTRs often contain both binding sites for microRNAs (miRNAs) as well as for regulatory proteins. By binding to specific sites within the 3'-UTR, miRNAs can decrease gene expression of various mRNAs by either inhibiting translation or directly causing degradation of the transcript. The 3'-UTR also may have silencer regions that bind repressor proteins that inhibit the expression of a mRNA.

The 3'-UTR often contains microRNA response elements (MREs). MREs are sequences to which miRNAs bind. These are prevalent motifs within 3'-UTRs. Among all regulatory motifs within the 3'-UTRs (e.g. including silencer regions), MREs make up about half of the motifs.

As of 2014, the miRBase web site, an archive of miRNA sequences and annotations, listed 28,645 entries in 233 biologic species. Of these, 1,881 miRNAs were in annotated human miRNA loci. miRNAs were predicted to have an average of about four hundred target mRNAs (affecting expression of several hundred genes). Freidman et al. estimate that >45,000 miRNA target sites within human mRNA 3'UTRs are conserved above background levels, and >60% of human protein-coding genes have been under selective pressure to maintain pairing to miRNAs.

Direct experiments show that a single miRNA can reduce the stability of hundreds of unique mRNAs. Other experiments show that a single miRNA may repress the production of hundreds of proteins, but that this repression often is relatively mild (less than 2-fold).

The effects of miRNA dysregulation of gene expression seem to be important in cancer. For instance, in gastrointestinal cancers, nine miRNAs have been identified as epigenetically altered and effective in down regulating DNA repair enzymes.

The effects of miRNA dysregulation of gene expression also seem to be important in neuropsychiatric disorders, such as schizophrenia, bipolar disorder, major depression, Parkinson's disease, Alzheimer's disease and autism spectrum disorders.

Translational Regulation

Direct regulation of translation is less prevalent than control of transcription or mRNA stability but is occasionally used. Inhibition of protein translation is a major target for toxins and antibiotics, so they can kill a cell by overriding its normal gene expression control. Protein synthesis inhibitors include the antibiotic neomycin and the toxin ricin.

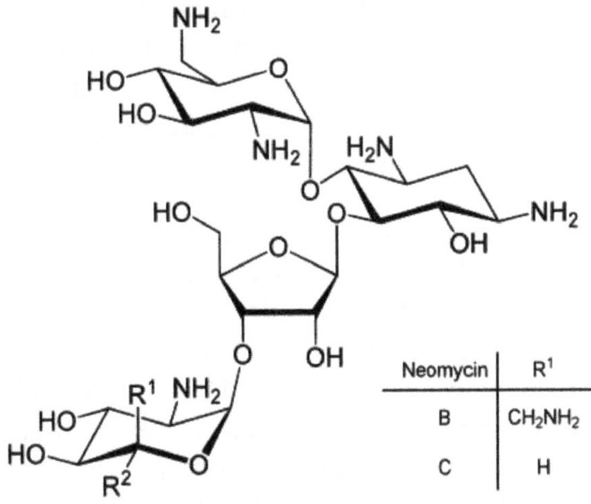

Neomycin	R^1	R^2
B	CH_2NH_2	H
C	H	CH_2NH_2

Neomycin is an example of a small molecule that reduces expression of all protein genes inevitably leading to cell death; it thus acts as an antibiotic.

Protein Degradation

Once protein synthesis is complete, the level of expression of that protein can be reduced by protein degradation. There are major protein degradation pathways in all prokaryotes and eukaryotes, of which the proteasome is a common component. An unneeded or damaged protein is often labeled for degradation by addition of ubiquitin.

Measurement

Measuring gene expression is an important part of many life sciences, as the ability to quantify the level at which a particular gene is expressed within a cell, tissue or organism can provide a lot of valuable information. For example, measuring gene expression can:

- Identify viral infection of a cell (viral protein expression).
- Determine an individual's susceptibility to cancer (oncogene expression).
- Find if a bacterium is resistant to penicillin (beta-lactamase expression).

Similarly, the analysis of the location of protein expression is a powerful tool, and this can be done on an organismal or cellular scale. Investigation of localization is particularly important for the study of development in multicellular organisms and as an indicator of protein function in single cells. Ideally, measurement of expression is done by detecting the final gene product (for many genes, this is the protein); however, it is often easier to detect one of the precursors, typically mRNA and to infer gene-expression levels from these measurements.

mRNA Quantification

Levels of mRNA can be quantitatively measured by northern blotting, which provides size and sequence information about the mRNA molecules. A sample of RNA is separated on an agarose gel and hybridized to a radioactively labeled RNA probe that is complementary to the target sequence. The radiolabeled RNA is then detected by an autoradiograph. Because the use of radioactive reagents makes the procedure time consuming and potentially dangerous, alternative labeling and detection methods, such as digoxigenin and biotin chemistries, have been developed. Perceived disadvantages of Northern blotting are that large quantities of RNA are required and that quantification may not be completely accurate, as it involves measuring band strength in an image of a gel. On the other hand, the additional mRNA size information from the Northern blot allows the discrimination of alternately spliced transcripts.

Another approach for measuring mRNA abundance is RT-qPCR. In this technique, reverse transcription is followed by quantitative PCR. Reverse transcription first generates a DNA template from the mRNA; this single-stranded template is called cDNA. The cDNA template is then amplified in the quantitative step, during which the fluorescence emitted by labeled hybridization probes or intercalating dyes changes as the DNA amplification process progresses. With a carefully constructed standard curve, qPCR can produce an absolute measurement of the number of copies of original mRNA, typically in units of copies per nanolitre of homogenized tissue or copies per cell. qPCR is very sensitive (detection of a single mRNA molecule is theoretically possible), but can be expensive depending on the type of reporter used; fluorescently labeled oligonucleotide probes are more expensive than non-specific intercalating fluorescent dyes.

For expression profiling, or high-throughput analysis of many genes within a sample, quantitative PCR may be performed for hundreds of genes simultaneously in the case of low-density arrays. A second approach is the hybridization microarray. A single array or "chip" may contain probes to determine transcript levels for every known gene in the genome of one or more organisms. Alternatively, "tag based" technologies like Serial analysis of gene expression (SAGE) and RNA-Seq, which can provide a relative measure of the cellular concentration of different mRNAs, can be used. An advantage of tag-based methods is the "open architecture", allowing for the exact measurement of any transcript, with a known or unknown sequence. Next-generation sequencing (NGS) such as RNA-Seq is another approach, producing vast quantities of sequence data that can be matched to a reference genome. Although NGS is comparatively time-consuming, expensive, and resource-intensive, it can identify single-nucleotide polymorphisms, splice-variants, and novel genes, and can also be used to profile expression in organisms for which little or no sequence information is available.

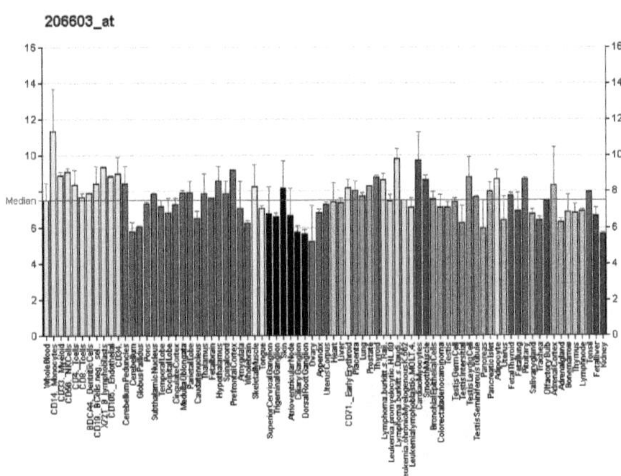

The RNA Expression profile of the GLUT4 Transporter (one of the main glucose transporters found in the human body.

Protein Quantification

For genes encoding proteins, the expression level can be directly assessed by a number of methods with some clear analogies to the techniques for mRNA quantification.

The most commonly used method is to perform a Western blot against the protein of interest—this gives information on the size of the protein in addition to its identity. A sample (often cellular lysate) is separated on a polyacrylamide gel, transferred to a membrane and then probed with an antibody to the protein of interest. The antibody can either be conjugated to a fluorophore or to horseradish peroxidase for imaging and/or quantification. The gel-based nature of this assay makes quantification less accurate, but it has the advantage of being able to identify later modifications to the protein, for example proteolysis or ubiquitination, from changes in size.

Localisation

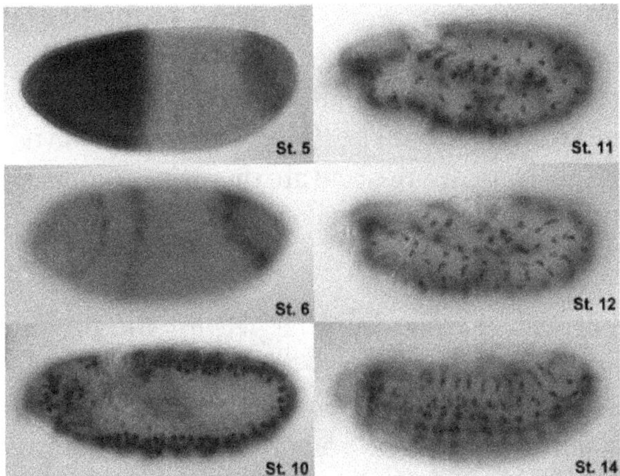

In situ-hybridization of Drosophila embryos at different developmental stages for the mRNA responsible for the expression of hunchback. High intensity of blue color marks places with high hunchback mRNA quantity.

Analysis of expression is not limited to quantification; localisation can also be determined. mRNA can be detected with a suitably labelled complementary mRNA strand and protein can be detected via labelled antibodies. The probed sample is then observed by microscopy to identify where the mRNA or protein is.

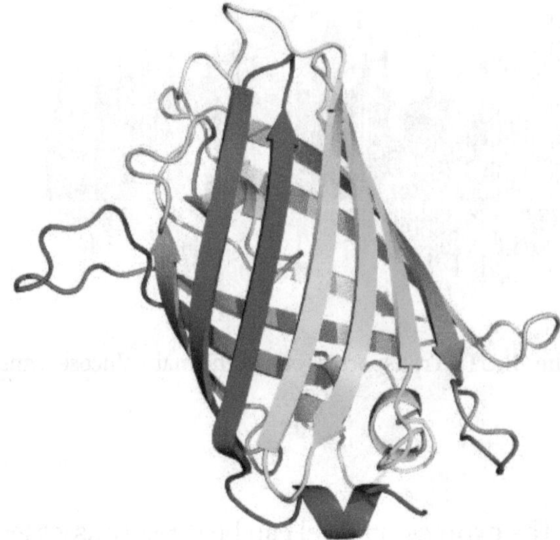

The three-dimensional structure of green fluorescent protein. The residues in the centre of the "barrel" are responsible for production of green light after exposing to higher energetic blue light. From PDB: 1EMA.

By replacing the gene with a new version fused to a green fluorescent protein (or similar) marker, expression may be directly quantified in live cells. This is done by imaging using a fluorescence microscope. It is very difficult to clone a GFP-fused protein into its native location in the genome without affecting expression levels so this method often cannot be used to measure endogenous gene expression. It is, however, widely used to measure the expression of a gene artificially introduced into the cell, for example via an expression vector. It is important to note that by fusing a target protein to a fluorescent reporter the protein's behavior, including its cellular localization and expression level, can be significantly changed.

The enzyme-linked immunosorbent assay works by using antibodies immobilised on a microtiter plate to capture proteins of interest from samples added to the well. Using a detection antibody conjugated to an enzyme or fluorophore the quantity of bound protein can be accurately measured by fluorometric or colourimetric detection. The detection process is very similar to that of a Western blot, but by avoiding the gel steps more accurate quantification can be achieved.

Expression System

An expression system is a system specifically designed for the production of a gene product of choice. This is normally a protein although may also be RNA, such as tRNA or a ribozyme. An expression system consists of a gene, normally encoded by DNA, and the molecular machinery required to transcribe the DNA into mRNA and translate the mRNA into protein using the reagents provided. In the broadest sense this includes every living cell but the term is more normally used to refer to expression as a laboratory tool. An expression system is therefore often artificial in some manner. Expression systems are, however, a fundamentally natural process. Viruses are an excellent example where they replicate by using the host cell as an expression system for the viral proteins and genome.

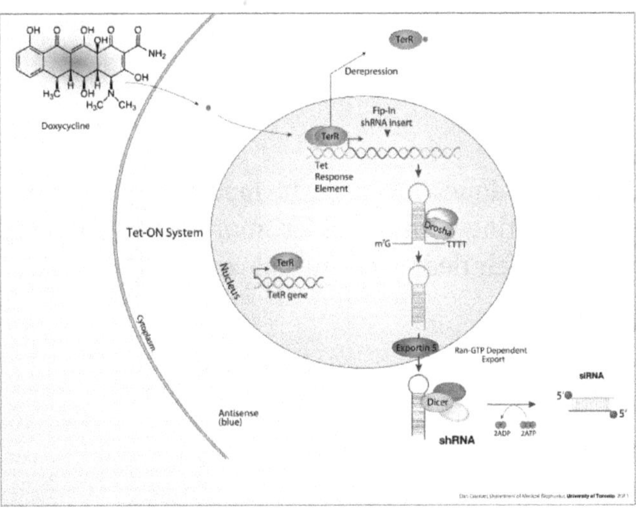

Tet-ON inducible shRNA system

Inducible Expression

Doxycycline is also used in "Tet-on" and "Tet-off" tetracycline controlled transcriptional activation to regulate transgene expression in organisms and cell cultures.

In Nature

In addition to these biological tools, certain naturally observed configurations of DNA (genes, promoters, enhancers, repressors) and the associated machinery itself are referred to as an expression system. This term is normally used in the case where a gene or set of genes is switched on under well defined conditions, for example, the simple repressor switch expression system in Lambda phage and the lac operator system in bacteria. Several natural expression systems are directly used or modified and used for artificial expression systems such as the Tet-on and Tet-off expression system.

Gene Networks

Genes have sometimes been regarded as nodes in a network, with inputs being proteins such as transcription factors, and outputs being the level of gene expression. The node itself performs a function, and the operation of these functions have been interpreted as performing a kind of information processing within cells and determines cellular behavior.

Gene networks can also be constructed without formulating an explicit causal model. This is often the case when assembling networks from large expression data sets. Covariation and correlation of expression is computed across a large sample of cases and measurements (often transcriptome or proteome data). The source of variation can be either experimental or natural (observational). There are several ways to construct gene expression networks, but one common approach is to compute a matrix of all pair-wise correlations of expression across conditions, time points, or individuals and convert the matrix (after thresholding at some cut-off value) into a graphical representation in which nodes represent genes, transcripts, or proteins and edges connecting these nodes represent the strength of association. Weighted correlation network analysis involves weighted networks defined by soft-thresholding the pairwise correlations among variables (e.g. measures of transcript abundance).

Techniques and Tools

The following experimental techniques are used to measure gene expression and are listed in roughly chronological order, starting with the older, more established technologies. They are divided into two groups based on their degree of multiplexity.

- Low-to-mid-plex techniques:
    - Reporter gene
    - Northern blot
    - Western blot
    - Fluorescent in situ hybridization
    - Reverse transcription PCR
- Higher-plex techniques:
    - SAGE
    - DNA microarray
    - Tiling array
    - RNA-Seq

Alternative Splicing

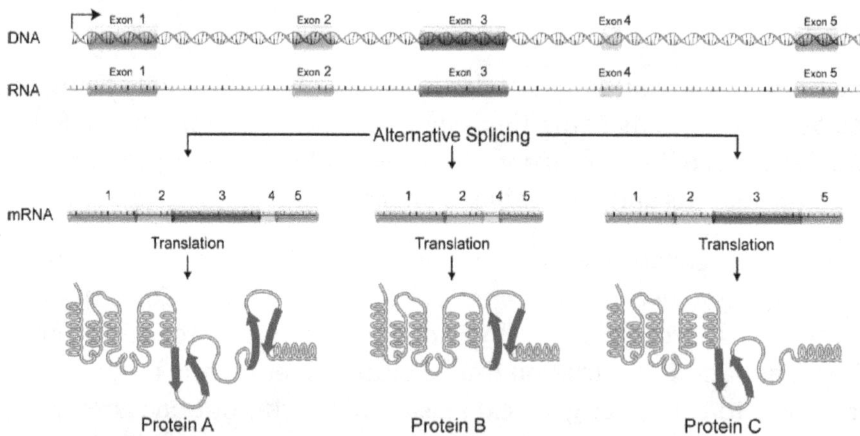

Alternative splicing produces three protein isoforms.

Alternative splicing, or differential splicing, is a regulated process during gene expression that results in a single gene coding for multiple proteins. In this process, particular exons of a gene may be included within or excluded from the final, processed messenger RNA (mRNA) produced from that gene. Consequently, the proteins translated from alternatively spliced mRNAs will contain differences in their amino acid sequence and, often, in their biological functions. Notably, alternative splicing allows the human genome to direct the synthesis of many more proteins than would be expected from its 20,000 protein-coding genes.

Alternative splicing occurs as a normal phenomenon in eukaryotes, where it greatly increases the biodiversity of proteins that can be encoded by the genome; in humans, ~95% of multi-exonic genes are alternatively spliced. There are numerous modes of alternative splicing observed, of which the most common is exon skipping. In this mode, a particular exon may be included in mRNAs under some conditions or in particular tissues, and omitted from the mRNA in others.

The production of alternatively spliced mRNAs is regulated by a system of trans-acting proteins that bind to cis-acting sites on the primary transcript itself. Such proteins include splicing activators that promote the usage of a particular splice site, and splicing repressors that reduce the usage of a particular site. Mechanisms of alternative splicing are highly variable, and new examples are constantly being found, particularly through the use of high-throughput techniques. Researchers hope to fully elucidate the regulatory systems involved in splicing, so that alternative splicing products from a given gene under particular conditions ("splicing variants") could be predicted by a "splicing code".

Abnormal variations in splicing are also implicated in disease; a large proportion of human genetic disorders result from splicing variants. Abnormal splicing variants are also thought to contribute to the development of cancer, and splicing factor genes are frequently mutated in different types of cancer.

Discovery

Alternative splicing was first observed in 1977. The Adenovirus produces five primary transcripts early in its infectious cycle, prior to viral DNA replication, and an additional one later, after DNA replication begins. The early primary transcripts continue to be produced after DNA replication begins. The additional primary transcript produced late in infection is large and comes from 5/6 of the 32kb adenovirus genome. This is much larger than any of the individual adenovirus mRNAs present in infected cells. Researchers found that the primary RNA transcript produced by adenovirus type 2 in the late phase was spliced in many different ways, resulting in mRNAs encoding different viral proteins. In addition, the primary transcript contained multiple polyadenylation sites, giving different 3' ends for the processed mRNAs.

In 1981, the first example of alternative splicing in a transcript from a normal, endogenous gene was characterized. The gene encoding the thyroid hormone calcitonin was found to be alternatively spliced in mammalian cells. The primary transcript from this gene contains 6 exons; the calcitonin mRNA contains exons 1–4, and terminates after a polyadenylation site in exon 4. Another mRNA is produced from this pre-mRNA by skipping exon 4, and includes exons 1–3, 5, and 6. It encodes a protein known as CGRP (calcitonin gene related peptide). Examples of alternative splicing in immunoglobin gene transcripts in mammals were also observed in the early 1980s.

Since then, alternative splicing has been found to be ubiquitous in eukaryotes. The "record-holder" for alternative splicing is a *D. melanogaster* gene called Dscam, which could potentially have 38,016 splice variants.

Modes

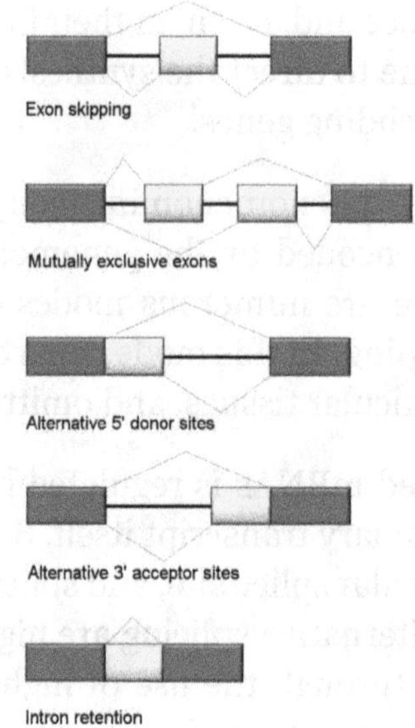

Traditional classification of basic types of alternative RNA splicing events.

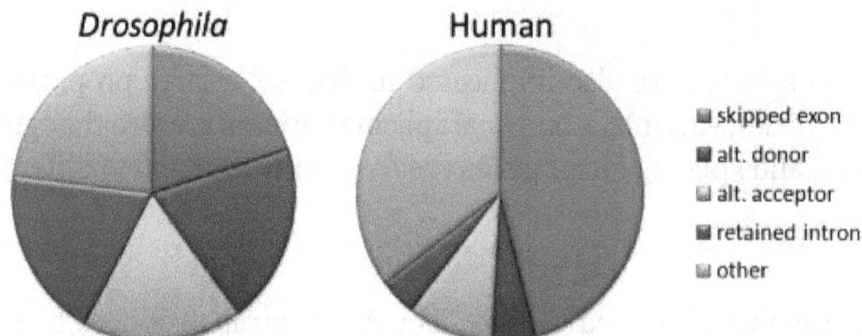

Relative frequencies of types of alternative splicing events differ between humans and fruit flies.

Five basic modes of alternative splicing are generally recognized.

- Exon skipping or cassette exon: in this case, an exon may be spliced out of the primary transcript or retained. This is the most common mode in mammalian pre-mRNAs.
- Mutually exclusive exons: One of two exons is retained in mRNAs after splicing, but not both.
- Alternative donor site: An alternative 5' splice junction (donor site) is used, changing the 3' boundary of the upstream exon.
- Alternative acceptor site: An alternative 3' splice junction (acceptor site) is used, changing the 5' boundary of the downstream exon.
- Intron retention: A sequence may be spliced out as an intron or simply retained. This is distinguished from exon skipping because the retained sequence is not flanked by introns. If the retained intron is in the coding region, the intron must encode amino acids in frame with the neighboring exons, or a stop codon or a shift in the reading frame will cause the protein to be non-functional. This is the rarest mode in mammals.

In addition to these primary modes of alternative splicing, there are two other main mechanisms by which different mRNAs may be generated from the same gene; multiple promoters and multiple polyadenylation sites. Use of multiple promoters is properly described as a transcriptional regulation mechanism rather than alternative splicing; by starting transcription at different points, transcripts with different 5'-most exons can be generated. At the other end, multiple polyadenylation sites provide different 3' end points for the transcript. Both of these mechanisms are found in combination with alternative splicing and provide additional variety in mRNAs derived from a gene.

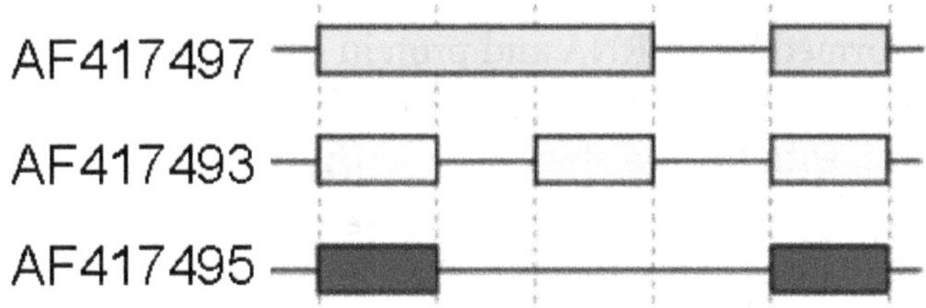

Schematic cutoff from 3 splicing structures in the murine hyaluronidase gene. Directionality of transcription from 5' to 3' is shown from left to right. Exons and introns are not drawn to scale.

These modes describe basic splicing mechanisms, but may be inadequate to describe complex splicing events. For instance, the figure to the right shows 3 spliceforms from the mouse hyaluronidase 3 gene. Comparing the exonic structure shown in the first line (green) with the one in the second line (yellow) shows intron retention, whereas the comparison between the second and the third spliceform (yellow vs. blue) exhibits exon skipping. A model nomenclature to uniquely designate all possible splicing patterns has recently been proposed.

Alternative Splicing Mechanisms

General Splicing Mechanism

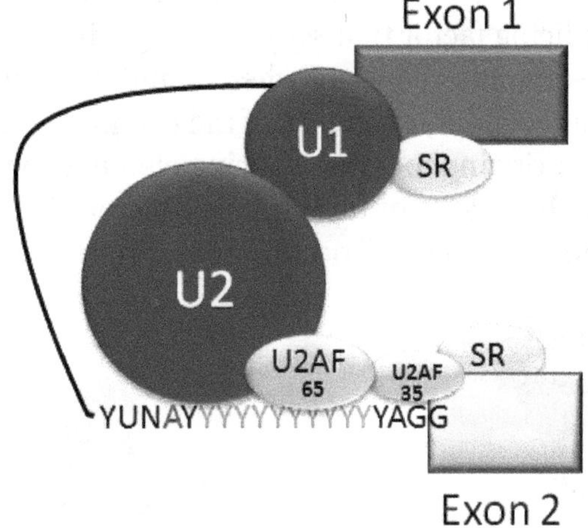

Spliceosome A complex defines the 5' and 3' ends of the intron before removal

When the pre-mRNA has been transcribed from the DNA, it includes several introns and exons. (In nematodes, the mean is 4–5 exons and introns; in the fruit fly *Drosophila* there can be more than 100 introns and exons in one transcribed pre-mRNA.) The exons to be retained in the mRNA are determined during the splicing process. The regulation and selection of splice sites are done by trans-acting splicing activator and splicing repressor proteins as well as cis-acting elements within the pre-mRNA itself such as exonic splicing enhancers and exonic splicing silencers.

The typical eukaryotic nuclear intron has consensus sequences defining important regions. Each intron has GU at its 5' end. Near the 3' end there is a branch site. The nucleotide at the branchpoint is always an A; the consensus around this sequence varies somewhat. In humans the branch site consensus sequence is yUnAy. The branch site is followed by a series of pyrimidines - the polypyrimidine tract - then by AG at the 3' end.

Splicing of mRNA is performed by an RNA and protein complex known as the spliceosome, containing snRNPs designated U1, U2, U4, U5, and U6 (U3 is not involved in mRNA splicing). U1 binds to the 5' GU and U2, with the assistance of the U2AF protein factors, binds to the branchpoint A within the branch site. The complex at this stage is known as the spliceosome A complex. Formation of the A complex is usually the key step in determining the ends of the intron to be spliced out, and defining the ends of the exon to be retained. (The U nomenclature derives from their high uridine content).

The U4,U5,U6 complex binds, and U6 replaces the U1 position. U1 and U4 leave. The remaining complex then performs two transesterification reactions. In the first transesterification, 5' end of the intron is cleaved from the upstream exon and joined to the branch site A by a 2',5'-phosphodiester linkage. In the second transesterification, the 3' end of the intron is cleaved from the downstream exon, and the two exons are joined by a phosphodiester bond. The intron is then released in lariat form and degraded.

Regulatory Elements and Proteins

Splicing is regulated by trans-acting proteins (repressors and activators) and corresponding cis-acting regulatory sites (silencers and enhancers) on the pre-mRNA. However, as part of the complexity of alternative splicing, it is noted that the effects of a splicing factor are frequently position-dependent. That is, a splicing factor that serves as a splicing activator when bound to an intronic enhancer element may serve as a repressor when bound to its splicing element in the context of an exon, and vice versa. The secondary structure of the pre-mRNA transcript also plays a role in regulating splicing, such as by bringing together splicing elements or by masking a sequence that would otherwise serve as a binding element for a splicing factor. Together, these elements form a "splicing code" that governs how splicing will occur under different cellular conditions.

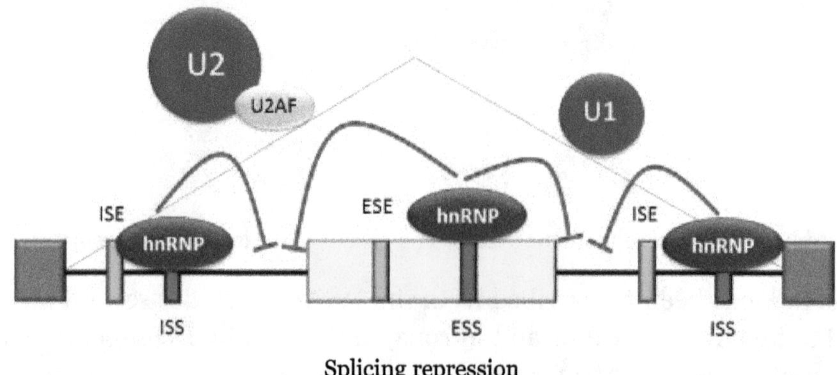

Splicing repression

There are two major types of cis-acting RNA sequence elements present in pre-mRNAs and they have corresponding trans-acting RNA-binding proteins. Splicing *silencers* are sites to which splicing repressor proteins bind, reducing the probability that a nearby site will be used as a splice junction. These can be located in the intron itself (intronic splicing silencers, ISS) or in a neighboring exon (exonic splicing silencers, ESS). They vary in sequence, as well as in the types of proteins that bind to them. The majority of splicing repressors are heterogeneous nuclear ribonucleoproteins (hnRNPs) such as hnRNPA1 and polypyrimidine tract binding protein (PTB). Splicing *enhancers* are sites to which splicing activator proteins bind, increasing the probability that a nearby site will be used as a splice junction. These also may occur in the intron (intronic splicing enhancers, ISE) or exon (exonic splicing enhancers, ESE). Most of the activator proteins that bind to ISEs and ESEs are members of the SR protein family. Such proteins contain RNA recognition motifs and arginine and serine-rich (RS) domains.

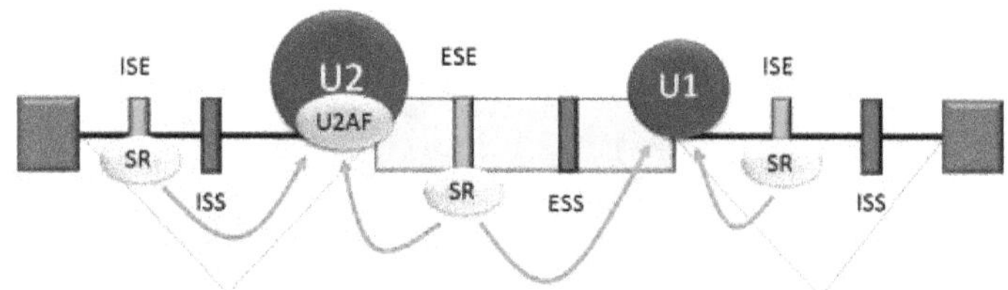

Splicing activation

In general, the determinants of splicing work in an inter-dependent manner that depends on context, so that the rules governing how splicing is regulated from a splicing code. The presence of a particular cis-acting RNA sequence element may increase the probability that a nearby site will be spliced in some cases, but decrease the probability in other cases, depending on context. The context within which regulatory elements act includes cis-acting context that is established by the presence of other RNA sequence features, and trans-acting context that is established by cellular conditions. For example, some cis-acting RNA sequence elements influence splicing only if multiple elements are present in the same region so as to establish context. As another example, a cis-acting element can have opposite effects on splicing, depending on which proteins are expressed in the cell (e.g., neuronal versus non-neuronal PTB). The adaptive significance of splicing silencers and enhancers is attested by studies showing that there is strong selection in human genes against mutations that produce new silencers or disrupt existing enhancers.

Examples

Exon Skipping: Drosophila dsx

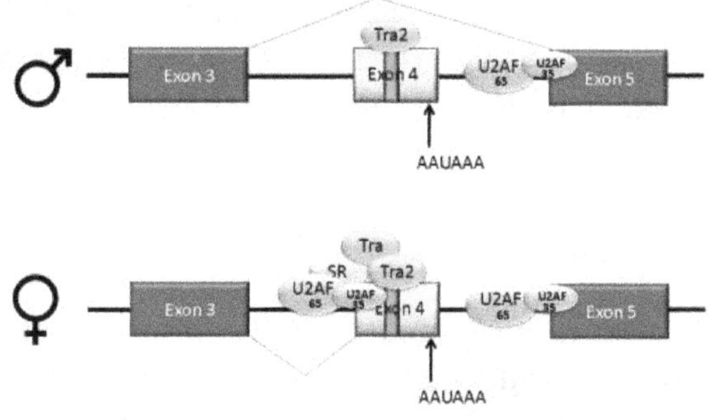

Alternative splicing of *dsx* pre-mRNA

Pre-mRNAs from the *D. melanogaster* gene *dsx* contain 6 exons. In males, exons 1,2,3,5,and 6 are joined to form the mRNA, which encodes a transcriptional regulatory protein required for male development. In females, exons 1,2,3, and 4 are joined, and a polyadenylation signal in exon 4 causes cleavage of the mRNA at that point. The resulting mRNA is a transcriptional regulatory protein required for female development.

This is an example of exon skipping. The intron upstream from exon 4 has a polypyrimidine tract that doesn't match the consensus sequence well, so that U2AF proteins bind poorly to it without assistance from splicing activators. This 3' splice acceptor site is therefore not used in males. Females, however, produce the splicing activator Transformer (Tra). The SR protein Tra2 is produced in both sexes and binds to an ESE in exon 4; if Tra is present, it binds to Tra2 and, along with another SR protein, forms a complex that assists U2AF proteins in binding to the weak polypyrimidine tract. U2 is recruited to the associated branchpoint, and this leads to inclusion of exon 4 in the mRNA.

Alternative Acceptor Sites: Drosophila Transformer

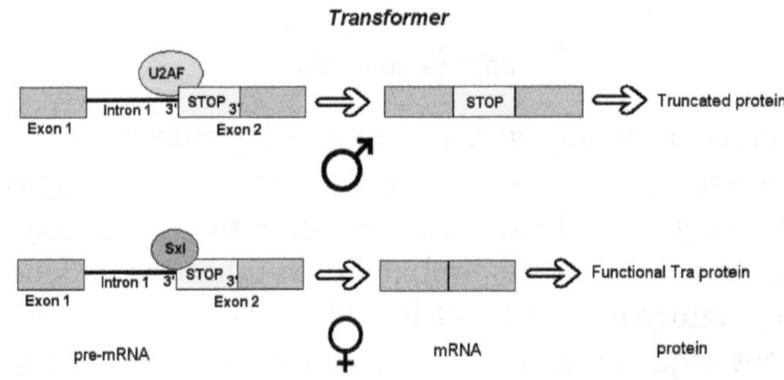

Alternative splicing of the *Drosophila Transformer* gene product.

Pre-mRNAs of the *Transformer* (Tra) gene of *Drosophila melanogaster* undergo alternative splicing via the alternative acceptor site mode. The gene Tra encodes a protein that is expressed only in females. The primary transcript of this gene contains an intron with two possible acceptor sites. In males, the upstream acceptor site is used. This causes a longer version of exon 2 to be included in the processed transcript, including an early stop codon. The resulting mRNA encodes a truncated protein product that is inactive. Females produce the master sex determination protein Sex lethal (Sxl). The Sxl protein is a splicing repressor that binds to an ISS in the RNA of the Tra transcript near the upstream acceptor site, preventing U2AF protein from binding to the polypyrimidine tract. This prevents the use of this junction, shifting the spliceosome binding to the downstream acceptor site. Splicing at this point bypasses the stop codon, which is excised as part of the intron. The resulting mRNA encodes an active Tra protein, which itself is a regulator of alternative splicing of other sex-related genes.

Exon Definition: Fas Receptor

Multiple isoforms of the Fas receptor protein are produced by alternative splicing. Two normally occurring isoforms in humans are produced by an exon-skipping mechanism. An mRNA including exon 6 encodes the membrane-bound form of the Fas receptor, which promotes apoptosis, or programmed cell death. Increased expression of Fas receptor in skin cells chronically exposed to the sun, and absence of expression in skin cancer cells, suggests that this mechanism may be important in elimination of pre-cancerous cells in humans. If exon 6 is skipped, the resulting mRNA encodes a soluble Fas protein that does not promote apoptosis. The inclusion or skipping of the exon depends on two antagonistic proteins, TIA-1 and polypyrimidine tract-binding protein (PTB).

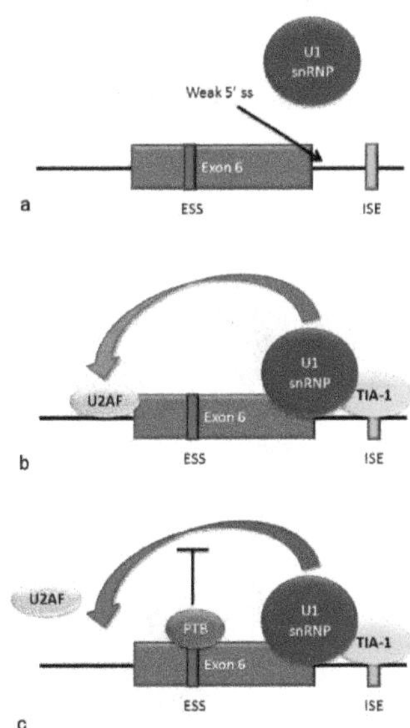

Alternative splicing of the Fas receptor pre-mRNA

- The 5' donor site in the intron downstream from exon 6 in the pre-mRNA has a weak agreement with the consensus sequence, and is not bound usually by the U1 snRNP. If U1 does not bind, the exon is skipped (see "a" in accompanying figure).
- Binding of TIA-1 protein to an intronic splicing enhancer site stabilizes binding of the U1 snRNP. The resulting 5' donor site complex assists in binding of the splicing factor U2AF to the 3' splice site upstream of the exon, through a mechanism that is not yet known (see b).
- Exon 6 contains a pyrimidine-rich exonic splicing silencer, *ure6*, where PTB can bind. If PTB binds, it inhibits the effect of the 5' donor complex on the binding of U2AF to the acceptor site, resulting in exon skipping (see c).

This mechanism is an example of exon definition in splicing. A spliceosome assembles on an intron, and the snRNP subunits fold the RNA so that the 5' and 3' ends of the intron are joined. However, recently studied examples such as this one show that there are also interactions between the ends of the exon. In this particular case, these exon definition interactions are necessary to allow the binding of core splicing factors prior to assembly of the spliceosomes on the two flanking introns.

Repressor-activator Competition: HIV-1 Tat Exon 2

HIV, the retrovirus that causes AIDS in humans, produces a single primary RNA transcript, which is alternatively spliced in multiple ways to produce over 40 different mRNAs. Equilibrium among differentially spliced transcripts provides multiple mRNAs encoding different products that are required for viral multiplication. One of the differentially spliced transcripts contains the *tat* gene, in which exon 2 is a cassette exon that may be skipped or included. The inclusion of tat exon 2 in the RNA is regulated by competition between the splicing repressor hnRNP A1 and the SR protein SC35. Within exon 2 an exonic splicing silencer sequence (ESS) and an exonic splicing enhancer sequence (ESE) overlap. If A1 repressor protein binds to the ESS, it initiates cooperative binding of multiple A1 molecules, extending into the 5' donor site upstream of exon 2 and preventing the binding of the core splicing factor U2AF35 to the polypyrimidine tract. If SC35 binds to the ESE, it prevents A1 binding and maintains the 5' donor site in an accessible state for assembly of the spliceosome. Competition between the activator and repressor ensures that both mRNA types (with and without exon 2) are produced.

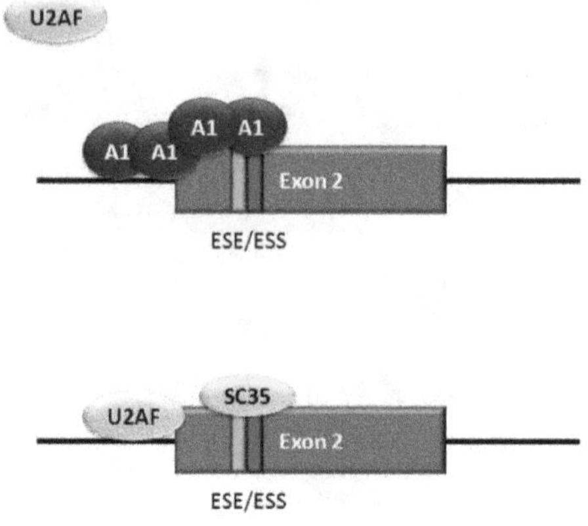

Alternative splicing of HIV-1 tat exon 2

Adaptive Significance

Alternative splicing is one of several exceptions to the original idea that one DNA sequence codes for one polypeptide (the One gene-one enzyme hypothesis). It might be more correct now to say "One gene – many polypeptides". External information is needed in order to decide which polypeptide is produced, given a DNA sequence and pre-mRNA. Since the methods of regulation are inherited, this provides novel ways for mutations to affect gene expression.

It has been proposed that for eukaryotes alternative splicing was a very important step towards higher efficiency, because information can be stored much more economically. Several proteins can be encoded by a single gene, rather than requiring a separate gene for each, and thus allowing a more varied proteome from a genome of limited size. It also provides evolutionary flexibility. A single point mutation may cause a given exon to be occasionally excluded or included from a transcript during splicing, allowing production of a new protein isoform without loss of the original protein. Studies have identified intrinsically disordered regions (see Intrinsically unstructured proteins) as enriched in the non-constitutive exons suggesting that protein isoforms may display functional diversity due to the alteration of functional modules within these regions. Such functional diversity achieved by isoforms is reflected by their expression patterns and can be predicted by machine learning approaches. Comparative studies indicate that alternative splicing preceded multicellularity in evolution, and suggest that this mechanism might have been co-opted to assist in the development of multicellular organisms.

Research based on the Human Genome Project and other genome sequencing has shown that humans have only about 30% more genes than the roundworm *Caenorhabditis elegans*, and only about twice as many as the fly *Drosophila melanogaster*. This finding led to speculation that the perceived greater complexity of humans, or vertebrates generally, might be due to higher rates of alternative splicing in humans than are found in invertebrates. However, a study on samples of 100,000 ESTs each from human, mouse, rat, cow, fly (*D. melanogaster*), worm (*C. elegans*), and the plant *Arabidopsis thaliana* found no large differences in frequency of alternatively spliced genes among humans and any of the other animals tested. Another study, however, proposed that these results were an artifact of the different numbers of ESTs available for the various organisms. When they compared alternative splicing frequencies in random subsets of genes from each organism, the authors concluded that vertebrates do have higher rates of alternative splicing than invertebrates.

Alternative Splicing and Disease

Changes in the RNA processing machinery may lead to mis-splicing of multiple transcripts, while single-nucleotide alterations in splice sites or cis-acting splicing regulatory sites may lead to differences in splicing of a single gene, and thus in the mRNA produced from a mutant gene's transcripts. A study in 2005 involving probabilistic analyses indicated that greater than 60% of human disease-causing mutations affect splicing rather than directly affecting coding sequences. A more recent study indicates that one-third of all hereditary diseases are likely to have a splicing component. Regardless of exact percentage, a number of splicing-related diseases do exist. As described below, a prominent example of splicing-related diseases is cancer.

Abnormally spliced mRNAs are also found in a high proportion of cancerous cells. Combined RNA-Seq and proteomics analyses have revealed striking differential expression of splice isoforms of key proteins in important cancer pathways. It is not always clear whether such aberrant patterns of splicing contribute to the cancerous growth, or are merely consequence of cellular abnormalities associated with cancer. Interestingly, for certain types of cancer, like in colorectal and prostate, the number of splicing errors per cancer has been shown to vary greatly between individual cancers, a phenomenon referred to as transcriptome instability. Transcriptome instability has further been shown to correlate grealty with reduced expression level of splicing factor genes. In fact, there is actually a reduction of alternative splicing in cancerous cells compared to normal ones, and the types of splicing differ; for instance, cancerous cells show higher levels of intron retention than normal cells, but lower levels of exon skipping. Some of the differences in splicing in cancerous cells may be due to the high frequency of somatic mutations in splicing factor genes, and some may result from changes in phosphorylation of trans-acting splicing factors. Others may be produced by changes in the relative amounts of splicing factors produced; for instance, breast cancer cells have been shown to have increased levels of the splicing factor SF2/ASF. One study found that a relatively small percentage (383 out of over 26000) of alternative splicing variants were significantly higher in frequency in tumor cells than normal cells, suggesting that there is a limited set of genes which, when mis-spliced, contribute to tumor development. It is believed however that the deleterious effects of mis-spliced transcripts are usually safeguarded and eliminated by a cellular posttranscriptional quality control mechanism termed Nonsense-mediated mRNA decay [NMD].

One example of a specific splicing variant associated with cancers is in one of the human DNMT genes. Three DNMT genes encode enzymes that add methyl groups to DNA, a modification that often has regulatory effects. Several abnormally spliced DNMT3B mRNAs are found in tumors and cancer cell lines. In two separate studies, expression of two of these abnormally spliced mRNAs in mammalian cells caused changes in the DNA methylation patterns in those cells. Cells with one of the abnormal mRNAs also grew twice as fast as control cells, indicating a direct contribution to tumor development by this product.

Another example is the *Ron* (*MST1R*) proto-oncogene. An important property of cancerous cells is their ability to move and invade normal tissue. Production of an abnormally spliced transcript of *Ron* has been found to be associated with increased levels of the SF2/ASF in breast cancer cells. The abnormal isoform of the Ron protein encoded by this mRNA leads to cell motility.

Overexpression of a truncated splice variant of the FOSB gene – ΔFosB – in a specific population of neurons in the nucleus accumbens has been identified as the causal mechanism involved in the induction and maintenance of an addiction to drugs and natural rewards.

Recent provocative studies point to a key function of chromatin structure and histone modifications in alternative splicing regulation. These insights suggest that epigenetic regulation determines not only what parts of the genome are expressed but also how they are spliced.

Genome-wide Analysis of Alternative Splicing

Genome-wide analysis of alternative splicing is a challenging task. Typically, alternatively spliced transcripts have been found by comparing EST sequences, but this requires sequencing of very large numbers of ESTs. Most EST libraries come from a very limited number of tissues, so tissue-specific splice variants are likely to be missed in any case. High-throughput approaches to investigate splicing have, however, been developed, such as: DNA microarray-based analyses, RNA-binding assays, and deep sequencing. These methods can be used to screen for polymorphisms or mutations in or around splicing elements that affect protein binding. When combined with splicing assays, including *in vivo* reporter gene assays, the functional effects of polymorphisms or mutations on the splicing of pre-mRNA transcripts can then be analyzed.

In microarray analysis, arrays of DNA fragments representing individual exons (*e.g.* Affymetrix exon microarray) or exon/exon boundaries (*e.g.* arrays from ExonHit or Jivan) have been used. The array is then probed with labeled cDNA from tissues of interest. The probe cDNAs bind to DNA from the exons that are included in mRNAs in their tissue of origin, or to DNA from the boundary where two exons have been joined. This can reveal the presence of particular alternatively spliced mRNAs.

CLIP (Cross-linking and immunoprecipitation) uses UV radiation to link proteins to RNA molecules in a tissue during splicing. A trans-acting splicing regulatory protein of interest is then precipitated using specific antibodies. When the RNA attached to that protein is isolated and cloned, it reveals the target sequences for that protein. Another method for identifying RNA-binding proteins and mapping their binding to pre-mRNA transcripts is "Microarray Evaluation of Genomic Aptamers by shift (MEGAshift)".net This method involves an adaptation of the "Systematic Evolution of Ligands by Exponential Enrichment (SELEX)" method together with a microarray-based readout. Use of the MEGAshift method has provided insights into the regulation of alternative splicing by allowing for the identification of sequences in pre-mRNA transcripts surrounding alternatively spliced exons that mediate binding to different splicing factors, such as ASF/SF2 and

PTB. This approach has also been used to aid in determining the relationship between RNA secondary structure and the binding of splicing factors.

Deep sequencing technologies have been used to conduct genome-wide analyses of mRNAs - unprocessed and processed - thus providing insights into alternative splicing. For example, results from use of deep sequencing indicate that, in humans, an estimated 95% of transcripts from multiexon genes undergo alternative splicing, with a number of pre-mRNA transcripts spliced in a tissue-specific manner. Functional genomics and computational approaches based on multiple instance learning have also been developed to integrate RNA-seq data to predict functions for alternatively spliced isoforms. Deep sequencing has also aided in the *in vivo* detection of the transient lariats that are released during splicing, the determination of branch site sequences, and the large-scale mapping of branchpoints in human pre-mRNA transcripts.

Use of reporter assays makes it possible to find the splicing proteins involved in a specific alternative splicing event by constructing reporter genes that will express one of two different fluorescent proteins depending on the splicing reaction that occurs. This method has been used to isolate mutants affecting splicing and thus to identify novel splicing regulatory proteins inactivated in those mutants.

Alternative Splicing Databases

There is a collection of alternative splicing databases. These databases are useful for finding genes having pre-mRNAs undergoing alternative splicing and alternative splicing events. Such as, the Plant Alternative Splicing Database contains alternative splicing information for several major cereal plant species and other plant species.

Essential Gene

Essential genes are those genes of an organism that are thought to be critical for its survival. However, being *essential* is highly dependent on the circumstances in which an organism lives. For instance, a gene required to digest starch is only essential if starch is the only source of energy. Recently, systematic attempts have been made to identify those genes that are absolutely required to maintain life, provided that all nutrients are available. Such experiments have led to the conclusion that the absolutely required number of genes for bacteria is on the order of about 250-300. These essential genes encode proteins to maintain a central metabolism, replicate DNA, translate genes into proteins, maintain a basic cellular structure, and mediate transport processes into and out of the cell. Most genes are not essential but convey selective advantages and increased fitness.

Bacteria: Genome-wide Studies

Two main strategies have been employed to identify essential genes on a genome-wide basis: directed deletion of genes and random mutagenesis using transposons. In the first case, individual genes (or ORFs) are completely deleted from the genome in a systematic way. In transposon-mediated mutagenesis transposons are randomly inserted in as many positions in a genome as possible, aiming to inactivate the targeted genes (see figure below). Insertion mutants that are still able to survive or grow are not in essential genes. A summary of such screens is shown in the table.

Organism	Mutagenesis	Method	Readout	ORFs	Non-ess.	Essen-tial	% Ess.	Notes
Mycoplasma genitalium/pneumoniae	Random	Population	Sequencing	482	130	265-350	55-73%	---
Mycoplasma genitalium	Random	Clones	Sequencing	482	100	382	79%	b,c
Staphylococcus aureus WCUH29	Random	Clones	Sequencing	2,600	n/a	168	n/a	b,c
Staphylococcus aureus RN4220	Random	Clones	Sequencing	2,892	n/a	658	23%	---
Haemophilus influenzae Rd	Random	Population	Foot-print-PCR	1,657	602	670	40%	---
Streptococcus pneumoniae Rx-1	Targeted	Clones	Colony formation	2,043	234	113	n/a	c
Streptococcus pneumoniae D39	Targeted	Clones	Colony formation	2,043	560	133	n/a	c
Streptococcus pyogenes 5448	Random	Trans-poson	Tn-seq	1,865	?	227	12%	---
Streptococcus pyogenes NZ131	Random	Trans-poson	Tn-seq	1,700	?	241	14%	---
Streptococcus sanguinis SK36	Targeted	Clones	Colony formation	2,270	2,052	218	10%	a
Mycobacterium tuberculosis H37Rv	Random	Population	Microarray	3,989	2,567	614	15%	---
Mycobacterium tuberculosis	Random	Trans-poson	?	3,989	?	401	10%	---
Mycobacterium tuberculosis H37Rv	Random	Trans-poson	NG-Se-quencing	3,989	?	774	19%	---
Mycobacterium tuberculosis	---	Computa-tional	Computa-tional	3,989	?	283	7%	---
Bacillus subtilis 168	Targeted	Clones	Colony formation	4,105	3,830	261	7%	a,d,g
Escherichia coli K-12 MG1655	Random	Population	Foot-print-PCR	4,308	3,126	620	14%	---
Escherichia coli K-12 MG1655	Targeted	Clones	Colony formation	4,308	2,001	n/a	n/a	a,e
Escherichia coli K-12 BW25113	Targeted	Clones	Colony formation	4,390	3,985	303	7%	a
Pseudomonas aeruginosa PAO1	Random	Clones	Sequencing	5,570	4,783	678	12%	a
Porphyromonas gingivalis	Random	Trans-poson	Sequencing	1,990	1,527	463	23%	---
Pseudomonas aeruginosa PA14	Random	Clones	Sequencing	5,688	4,469	335	6%	a,f
Salmonella typhimurium	Random	Clones	Sequencing	4,425	n/a	257	~11%	b,c

Helicobacter pylori G27	Random	Population	Microarray	1,576	1,178	344	22%	---
Campylobacter jejuni	Random	Population	Microarray	1,654	?	195	12%	---
Corynebacterium glutamicum	Random	Population	?	3,002	2,352	650	22%	---
Francisella novicida	Random	Transposon	?	1,719	1,327	392	23%	---
Mycoplasma pulmonis UAB CTIP	Random	Transposon	?	782	472	310	40%	---
Vibrio cholerae N16961	Random	Transposon	?	3,890	?	779	20%	---
Salmonella Typhi	Random	Transposon	?	4,646	?	353	8%	---
Staphylococcus aureus	Random	Transposon	?	~2,600	?	351	14%	---
Caulobacter crescentus	Random	Transposon	?	3,767	?	480	13%	---
Neisseria meningitidis	Random	Transposon	?	2,158	?	585	27%	---
Desulfovibrio alaskensis	Random	Transposon	Sequencing	3,258	2,871	387	12%	---

Table 1. Essential genes in bacteria. Mutagenesis: *targeted* mutants are gene deletions; *random* mutants are transposon insertions. Methods: *Clones* indicate single gene deletions, *population* indicates whole population mutagenesis, e.g. using transposons. Essential genes from population screens include genes essential for fitness. ORFs: number of all open reading frames in that genome. Notes: (a) mutant collection available; (b) direct essentiality screening method (e.g. via antisense RNA) that does not provide information about nonessential genes. (c) Only partial dataset available. (d) Includes predicted gene essentiality and data compilation from published single-gene essentiality studies. (e) Project in progress. (f) Deduced by comparison of the two gene essentiality datasets obtained independently in the *P. aeruginosa* strains PA14 and PAO1. (g) The original result of 271 essential genes has been corrected to 261, with 31 genes that were thought to be essential being in fact non-essential whereas 20 novel essential genes have been described since then.

Essential genes in *Mycobacterium tuberculosis* H37Rv as found by using transposons which insert in random positions in the genome. If no transposons are found in a gene, the gene is most likely essential as it cannot tolerate any insertion. In this example, essential heme biosynthetic genes *hemA, hemB, hemC, hemD* are devoid of insertions. The number of sequence reads ("reads/TA") is shown for the indicated region of the H37Rv chromosome. Potential TA dinucleotide insertions sites are indicated. Image from Griffin et al. 2011.

Eukaryotes

Yeasts are the only eukaryotic species in which systematic and "complete" essentiality screens have been carried out. In *Saccharomyces cerevisiae* (budding yeast) 15-20% of all genes are essential. In *Schizosaccharomyces pombe* (fission yeast) 4,836 heterozygous deletions covering 98.4% of the 4,914 protein coding open reading frames have been constructed. 1,260 of these deletions turned out to be essential.

Similar screens are more difficult to carry out in other multicellular organisms, including mammals (as a model for humans), due to technical reasons, and their results are less clear. However, various methods have been developed for the nematode worm *C. elegans*, the fruit fly, and zebrafish (see table). A recent study of 900 mouse genes concluded that 42% of them were essential although the selected genes were not representative. A summary of essentiality screens is shown in the table below (mostly based on the Database of Essential Genes).

Organism	Method	Essential genes
Arabidopsis thaliana	T-DNA insertion	777
Caenorhabditis elegans (worm)	RNA interference	294
Danio rerio (zebrafish)	Insertion mutagenesis	288
Drosophila melanogaster (fruit fly)	P-element insertion mutagenesis	339
Homo sapiens (human)	Literature search	118
Homo sapiens (human)	mouse orthologs	2,472
Mus musculus (mouse)	Literature search	2114
Saccharomyces cerevisiae (yeast)	Single-gene deletions	878
Saccharomyces cerevisiae (yeast)	Single-gene deletions	1,105
Schizosaccharomyces pombe (yeast)	Single-gene deletions	1,260

Human

Gene knockout experiments are not possible or at least not ethical in humans. However, natural mutations have led to the identification of mutations that lead to early embryonic or later death. Note that many genes in humans are not absolutely essential for survival but can cause severe disease when mutated. Such mutations are catalogued in the Online Mendelian Inheritance in Man (OMIM) database. In a computational analysis of genetic variation and mutations in 2,472 human orthologs of known essential genes in the mouse, Georgi et al. found strong, purifying selection and comparatively reduced levels of sequence variation, indicating that these human genes are essential too.

Non-essential genes in humans. While it may be difficult to prove that a gene is essential in humans, it can be demonstrated that a gene is not essential or not even causing disease. For instance, sequencing the genomes of 2,636 Icelandic citizens and the genotyping of 101,584 additional subjects found 8,041 individuals who had 1 gene completely knocked out (i.e. these people were homozygous for a non-functional gene). Of the 8,041 individuals with complete knock-outs, 6,885 were estimated to be homozygotes, 1,249 were estimated to be compound heterozygotes (i.e. they had both alleles of a gene knocked out but the two alleles had different mutations). In these individuals, a total of 1,171 of the 19,135 human (RefSeq) genes (6.1%) were completely knocked out. It was concluded that these 1,171 genes are *non-essential* in humans — at least no associated diseases were reported. Similarly, the exome sequences of 3222 British Pakistani-heritage adults with high parental relatedness revealed 1111 rare-variant homozygous genotypes with predicted loss of gene function (LOF = knockouts) in 781 genes. This study found an average of 140 predicted LOF genotypes (per subject), including 16 rare (minor allele frequency <1%) heterozygotes, 0.34 rare homozygotes, 83.2 common heterozygotes and 40.6 common homozygotes. Nearly all rare homozygous LOF genotypes were found within autozygous segments (94.9%). Even though most of these individuals had no obvious health issue arising from their defective genes, it is possible that minor health issues may be found upon more detailed examination.

Viruses

Screens for essential genes have been carried out in a few viruses. For instance, human cytomegalovirus (CMV) was found to have 41 essential, 88 nonessential, and 27 augmenting ORFs (150 total ORFs). Most essential and augmenting genes are located in the central region, and nonessential genes generally cluster near the ends of the viral genome.

Tscharke and Dobson (2015) compiled a comprehensive survey of essential genes in Vaccinia Virus and assigned roles to each of the 223 ORFs of the Western Reserve (WR) strain and 207 ORFs of the Copenhagen strain, assessing their role in replication in cell culture. According to their definition, a gene is considered essential (i.e. has a role in cell culture) if its deletion results in a decrease in virus titre of greater than 10-fold in either a single or multiple step growth curve. All genes involved in wrapped virion production, actin tail formation, and extracellular virion release were also considered as essential. Genes that influence plaque size, but not replication were defined as non-essential. By this definition 93 genes are required for Vaccinia Virus replication in cell culture, while 108 and 94 ORFs, from WR and Copenhagen respectively, are non-essential. Vaccinia viruses with deletions at either end of the genome behaved as expected, exhibiting only mild or host range defects. In contrast, combining deletions at both ends of the genome for VACV strain WR caused a devastating growth defect on all cell lines tested. This demonstrates that single gene deletions are not sufficient to assess the essentiality of genes and that more genes are essential in Vaccinia virus than originally thought.

One of the bacteriophages screened for essential genes includes mycobacteriophage Giles. At least 35 of the 78 predicted Giles genes (45%) are non-essential for lytic growth. 20 genes were found to be essential. A major problem with phage genes is that a majority of their genes remain functionally unknown, hence their role is difficult to assess. A screen of *Salmonella enterica* phage SPN3US revealed 13 essential genes although it remains a bit obscure how many genes were really tested.

Quantitative Gene Essentiality Analysis

Most genes are neither absolutely essential nor absolutely non-essential. Ideally their contribution to cell or organismal growth needs to be measured quantitatively, e.g. by determining how much growth rate is reduced in a mutant compared to "wild-type" (which may have been chosen arbitrarily from a population). For instance, a particular gene deletion may reduce growth rate (or fertility rate or other characters) to 90% of the wild-type.

Synthetic Lethality

Two genes are synthetic lethal if neither one is essential but when both are mutated the double-mutant is lethal. Some studies have estimated that the number of synthetic lethal genes may be on the order of 45% of all genes.

Conditionally Essential Genes

Many genes are essential only under certain circumstances. For instance, if the amino acid lysine is supplied to a cell any gene that is required to make lysine is non-essential. However, when there is no lysine supplied, genes encoding enzymes for lysine biosynthesis become essential, as no protein synthesis is possible without lysine.

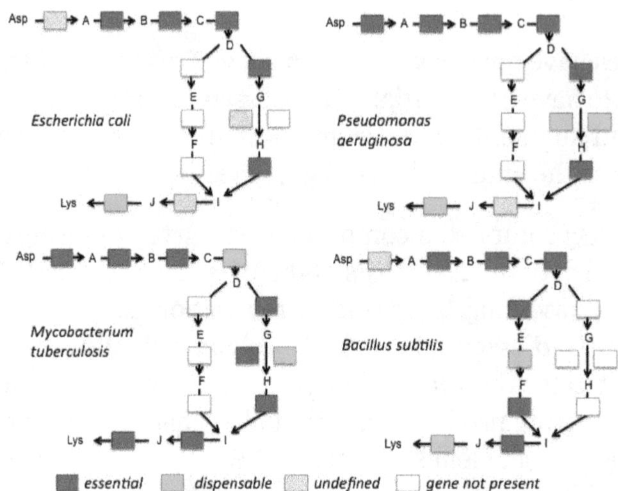

A schematic view of essential genes (or proteins) in lysine biosynthesis of different bacteria. The same protein may be essential in one species but not another.

Streptococcus pneumoniae appears to require 147 genes for growth and survival in saliva, more than the 113-133 that have been found in previous studies.

The deletion of a gene may result in death or in a block of cell division. While the latter case may implicate "survival" for some time, without cell division the cell may still die eventually. Similarly, instead of blocked cell division a cell may have reduced growth or metabolism ranging from nearly undetectable to almost normal. Thus, there is gradient from "essential" to completely non-essential, again depending on the condition. Some authors have thus distinguished between genes "essential for survival" and "essential for fitness".

The role of genetic background. Similar to environmental conditions, the genetic background can determine the essentiality of a gene: a gene may be essential in one individual but not another, given his or her genetic background. Gene duplications are one possible explanation (see below).

Essential Genes and Gene Duplications

Many genes are duplicated within a genome. Such duplications (paralogs) often render essential genes non-essential because the duplicate can replace the original copy. For instance, the gene encoding the enzyme aspartokinase is essential in *E. coli*. By contrast, the *Bacillus subtilis* genome contains three copies of this gene, none of which is essential on its own. However, a triple-deletion of all three genes is lethal. In such cases, the essentiality of a gene or a group of paralogs can often be predicted based on the essentiality of an essential single gene in a different species. In yeast, few of the essential genes are duplicated within the genome: 8.5% of the non-essential genes, but only 1% of the essential genes have a homologue in the yeast genome.

In the worm *C. elegans*, non-essential genes are highly over-represented among duplicates, possibly because duplication of essential genes causes overexpression of these genes. Woods et al. found that non-essential genes are more often successfully duplicated (fixed) and lost compared to essential genes. By contrast, essential genes are less often duplicated but upon successful duplication are maintained over longer periods.

Conservation of Essential Genes

In bacteria, essential genes appear to be more conserved than nonessential genes but the correlation is not very strong. For instance, only 34% of the *B. subtilis* essential genes have reliable orthologs in all Firmicutes and 61% of the *E. coli* essential genes have reliable orthologs in all Gamma-proteobacteria. Fang et al. (2005) defined persistent genes as the genes present in more than 85% of the genomes of the clade. They found 475 and 611 of such genes for *B. subtilis* and *E. coli*, respectively. Furthermore, they classified genes into five classes according to persistence and essentiality: persistent genes, essential genes, persistent nonessential (PNE) genes (276 in *B. subtilis*, 409 in *E. coli*), essential nonpersistent (ENP) genes (73 in *B. subtilis*, 33 in *E. coli*), and nonpersistent nonessential (NPNE) genes (3,558 in *B. subtilis*, 3,525 in *E. coli*). Fang et al. found 257 persistent genes, which exist both in *B. subtilis* (for the Firmicutes) and *E. coli* (for the Gamma-proteobacteria). Among these, 144 (respectively 139) were previously identified as essential in *B. subtilis* (respectively *E. coli*) and 25 (respectively 18) of the 257 genes are not present in the 475 *B. subtilis* (respectively 611 *E. coli*) persistent genes. All the other members of the pool are PNE genes.

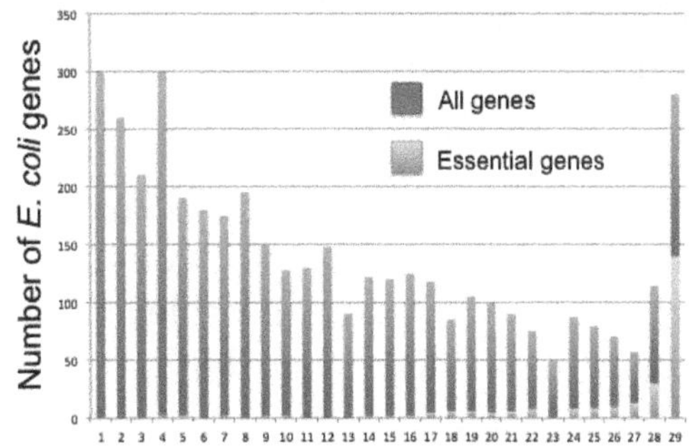

Conservation of essential genes in bacteria, adapted from

In eukaryotes, 83% of the one-to-one orthologs between *Schizosaccharomyces pombe* and *Saccharomyces cerevisiae* have conserved essentiality, that is, they are nonessential in both species or essential in both species. The remaining 17% of genes are nonessential in one species and essential in the other. This is quite remarkable, given that *S. pombe* is separated from *S. cerevisiae* by approximately 400 million years of evolution.

In general, highly conserved and thus older genes (i.e. genes with earlier phylogenetic origin) are more likely to be essential than younger genes - even if they have been duplicated.

Studying Essential Genes

The experimental study of essential genes is limited by the fact that, by definition, inactivation of an essential gene is lethal to the organism. Therefore they can not be simply deleted or mutated to analyze the resulting phenotypes (a common technique in genetics).

There are, however, some circumstances in which essential genes can be manipulated. In diploid organisms, only a single functional copy of some essential genes may be needed (haplosufficiency), with the heterozygote displaying an instructive phenotype. Some essential genes can tolerate mutations that are deleterious, but not wholly lethal, since they do not completely abolish the gene's function.

Computational analysis can reveal many properties of proteins without analyzing them experimentally, e.g. by looking at homologous proteins, function, structure etc. The products of essential genes can also be in studied when expressed in other organisms, or when purified and studied *in vitro*.

Conditionally essential genes are easier to study. Temperature-sensitive variants of essential genes have been identified which encode products that lose function at high temperatures, and so only show a phenotype at increased temperature.

Reproducibility

If screens for essential genes are repeated in independent laboratories, they often result in different gene lists. For instance, screens in *E. coli* have yielded from ~300 to ~600 essential genes Such differences are even more pronounced when different bacterial strains are used. A common explanation is that the experimental conditions are different or that the nature of the mutation may be different (e.g. a complete gene deletion vs. a transposon mutant). Transposon screens in particular are hard to reproduce, given that a transposon can insert at many positions within a gene. Insertions towards the 3' end of an essential gene may not have a lethal phenotype (or no phenotype at all) and thus may not be recognized as such. This can lead to erroneous annotations (here: false negatives).

Comparison of CRISPR/cas9 and RNAi screens. Screens to identify essential genes in the human chronic myelogenous leukemia cell line K562 with these two methods showed only limited overlap. At a 10% false positive rate there were ~4,500 genes identified in the Cas9 screen versus ~3,100 in the shRNA screen, with only ~1,200 genes identified in both.

Different Genes are Essential in Different Organisms

Different organisms have different essential genes. For instance, *Bacillus subtilis* has 271 essential genes. About one-half (150) of the orthologous genes in *E. coli* are also essential. Another 67 genes that are essential in *E. coli* are not essential in *B. subtilis*, while 86 *E. coli* essential genes have no *B. subtilis* ortholog.

In *Mycoplasma genitalium* at least 18 genes are essential that are not essential in *M. bovis*.

Predicting Essential Genes

Essential genes can be predicted computationally. However, most methods use experimental data ("training sets") to some extent. Chen et al. determined four criteria to select training sets for such predictions: (1) essential genes in the selected training set should be reliable; (2) the growth conditions in which essential genes are defined should be consistent in training and prediction sets; (3) species used as training set should be closely related to the target organism; and (4) organisms used as training and prediction sets should exhibit similar phenotypes or lifestyles. They also found that the size of the training set should be at least 10% of the total genes to yield accurate predictions. Some approaches for predicting essential genes are:

Comparative genomics. Shortly after the first genomes (of *Haemophilus influenzae* and *Mycoplasma genitalium*) became available, Mushegian et al. tried to predict the number of essential genes based on common genes in these two species. It was surmised that only essential genes should be conserved over the long evolutionary distance that separated the two bacteria. This study identified approximately 250 candidate essential genes. As more genomes became available the number of predicted essential genes kept shrinking because more genomes shared fewer and fewer genes. As a consequence, it was concluded that the universal conserved core consists of less than 40 genes. However, this set of conserved genes is not identical to the set of essential genes as different species rely on different essential genes.

A similar approach has been used to infer essential genes from the pan-genome of *Brucella* species. 42 complete *Brucella* genomes and a total of 132,143 protein-coding genes were used to predict 1252 potential essential genes, derived from the core genome by comparison with a prokaryote database of essential genes.

Hua et al. used Machine Learning to predict essential genes in 25 bacterial species.

Hurst index. Liu et al. (2015) used the Hurst exponent, a characteristic parameter to describe long-range correlation in DNA to predict essential genes. In 31 out of 33 bacterial genomes the significance levels of the Hurst exponents of the essential genes were significantly higher than for the corresponding full-gene-set, whereas the significance levels of the Hurst exponents of the non-essential genes remained unchanged or increased only slightly.

Minimal genomes. It was also thought that essential genes could be inferred from minimal genomes which supposedly contain only essential genes. The problem here is that the smallest genomes belong to parasitic (or symbiontic) species which can survive with a reduced gene set as they obtain many nutrients from their hosts. For instance, one of the smallest genomes is that of *Hodgkinia cicadicola*, a symbiont of cicadas, containing only 144 Kb of DNA encoding only 188 genes. Like other symbionts, *Hodgkinia* receives many of its nutrients from its host, so its genes do not need to be essential.

Metabolic modelling. Essential genes may be also predicted in completely sequenced genomes by metabolic reconstruction, that is, by reconstructing the complete metabolism from the gene content and then identifying those genes and pathways that have been found to be essential in other species. However, this method can be compromised by proteins of unknown function. In addition, many organisms have backup or alternative pathways which have to be taken into account. Metabolic modeling was also used by Basler (2015) to develop a method to predict essential metabolic genes. Flux balance analysis, a method of metabolic modeling, has recently been used to predict essential genes in clear cell renal cell carcinoma metabolism.

Genes of unknown function. Surprisingly, a significant number of essential genes has no known function. For instance, among the 385 essential candidates in *M. genitalium*, no function could be ascribed to 95 genes even though this number had been reduced to 75 by 2011.

ZUPLS. Song et al. presented a novel method to predict essential genes that only uses the Z-curve and other sequence-based features. Such features can be calculated readily from the DNA/amino acid sequences. However, the reliability of this method remains a bit obscure.

Essential gene prediction servers. Guo et al. (2015) have developed three online services to predict essential genes in bacterial genomes. These freely available tools are applicable for single gene sequences without annotated functions, single genes with definite names, and complete genomes of bacterial strains.

Essential Protein Domains

Although most essential genes encode proteins, many essential proteins consist of a single domain. This fact has been used to identify essential protein domains. Goodacre et al. have identified hundreds of essential domains of unknown function (eDUFs). Lu et al. presented a similar approach and identified 3,450 domains that are essential in at least one microbial species.

Mendelian Inheritance

Mendelian inheritance is a type of biological inheritance that follows the laws originally proposed by Gregor Mendel in 1865 and 1866 and re-discovered in 1900. These laws were initially very controversial. When Mendel's theories were integrated with the Boveri–Sutton chromosome theory of inheritance by Thomas Hunt Morgan in 1915, they became the core of classical genetics. Ronald Fisher later combined these ideas with the theory of natural selection in his 1930 book *The Genetical Theory of Natural Selection*, putting evolution onto a mathematical footing and forming the basis for population genetics and the modern evolutionary synthesis.

Gregor Mendel, the German-speaking Augustinian monk who founded the modern science of genetics.

History

The principles of Mendelian inheritance were named for and first derived by Gregor Johann Mendel, a nineteenth-century Austrian monk who formulated his ideas after conducting simple hybridization experiments with pea plants (*Pisum sativum*) he had planted in the garden of his monastery. Between 1856 and 1863, Mendel cultivated and tested some 5,000 pea plants. From these experiments, he induced two generalizations which later became known as *Mendel's Principles of Heredity* or *Mendelian inheritance*. He described these principles in a two-part paper, *Versuche über Pflanzen-Hybriden* (*Experiments on Plant Hybridization*), that he read to the Natural History Society of Brno on 8 February and 8 March 1865, and which was published in 1866.

Mendel's conclusions were largely ignored. Although they were not completely unknown to biologists of the time, they were not seen as generally applicable, even by Mendel himself, who thought they only applied to certain categories of species or traits. A major block to understanding their significance was the importance attached by 19th-century biologists to the apparent blending of inherited traits in the overall appearance of the progeny, now known to be due to multi-gene interactions, in contrast to the organ-specific binary characters studied by Mendel. In 1900, however, his work was "re-discovered" by three European scientists, Hugo de Vries, Carl Correns, and Erich von Tschermak. The exact nature of the "re-discovery" has been somewhat debated: De Vries published first on the subject, mentioning Mendel in a footnote, while Correns pointed out Mendel's priority after having read De Vries' paper and realizing that he himself did not have priority. De Vries may not have acknowledged truthfully how much of his knowledge of the laws came from his own work and how much came only after reading Mendel's paper. Later scholars have accused Von Tschermak of not truly understanding the results at all.

Regardless, the "re-discovery" made Mendelism an important but controversial theory. Its most vigorous promoter in Europe was William Bateson, who coined the terms "genetics" and "allele" to describe many of its tenets. The model of heredity was highly contested by other biologists because it implied that heredity was discontinuous, in opposition to the apparently continuous variation observable for many traits. Many biologists also dismissed the theory because they were not sure it would apply to all species. However, later work by biologists and statisticians such as Ronald Fisher showed that if multiple Mendelian factors were involved in the expression of an individual trait, they could produce the diverse results observed, and thus showed that Mendelian genetics is compatible with natural selection. Thomas Hunt Morgan and his assistants later integrated Mendel's theoretical model with the chromosome theory of inheritance, in which the chromosomes of cells were thought to hold the actual hereditary material, and created what is now known as classical genetics, a highly successful foundation which eventually cemented Mendel's place in history.

Mendel's findings allowed scientists such as Fisher and J.B.S. Haldane to predict the expression of traits on the basis of mathematical probabilities. An important aspect of Mendel's success can be traced to his decision to start his crosses only with plants he demonstrated were true-breeding. He also only measured absolute (binary) characteristics, such as color, shape, and position of the seeds, rather than quantitatively variable characteristics. He expressed his results numerically and subjected them to statistical analysis. His method of data analysis and his large sample size gave credibility to his data. He also had the foresight to follow several successive generations (F2, F3) of pea plants and record their variations. Finally, he performed "test crosses" (backcrossing descendants of the initial hybridization to the initial true-breeding lines) to reveal the presence and proportions of recessive characters.

Mendel's Laws

Mendel discovered that, when he crossed purebred white flower and purple flower pea plants (the parental or P generation), the result was not a blend. Rather than being a mix of the two, the offspring (known as the F_1 generation) was purple-flowered. When Mendel self-fertilized the F_1 generation pea plants, he obtained a purple flower to white flower ratio in the F_2 generation of 3 to 1. The results of this cross are tabulated in the Punnett square to the right.

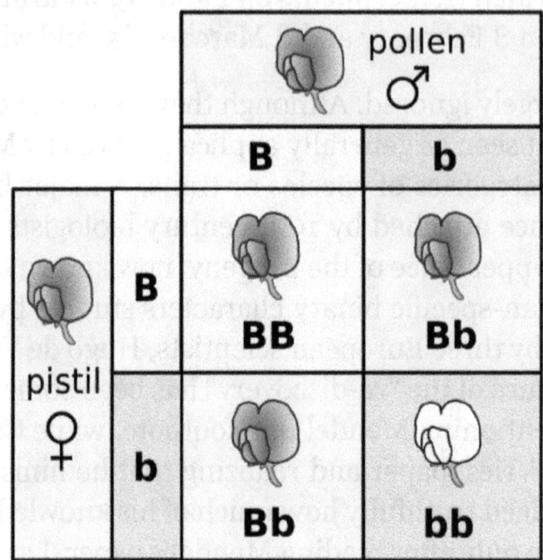

A Punnett square for one of Mendel's pea plant experiments.

He then conceived the idea of heredity units, which he called "factors". Mendel found that there are alternative forms of factors—now called genes—that account for variations in inherited characteristics. For example, the gene for flower color in pea plants exists in two forms, one for purple and the other for white. The alternative "forms" are now called alleles. For each biological trait, an organism inherits two alleles, one from each parent. These alleles may be the same or different. An organism that has two identical alleles for a gene is said to be homozygous for that gene (and is called a homozygote). An organism that has two different alleles for a gene is said be heterozygous for that gene (and is called a heterozygote).

Mendel also hypothesized that allele pairs separate randomly, or segregate, from each other during the production of gametes: egg and sperm. Because allele pairs separate during gamete production, a sperm or egg carries only one allele for each inherited trait. When sperm and egg unite at fertilization, each contributes its allele, restoring the paired condition in the offspring. This is called the Law of Segregation. Mendel also found that each pair of alleles segregates independently of the other pairs of alleles during gamete formation. This is known as the Law of Independent Assortment.

The genotype of an individual is made up of the many alleles it possesses. An individual's physical appearance, or phenotype, is determined by its alleles as well as by its environment. The presence of an allele does not mean that the trait will be expressed in the individual that possesses it. If the two alleles of an inherited pair differ (the heterozygous condition), then one determines the organism's appearance and is called the dominant allele; the other has no noticeable effect on the organism's appearance and is called the recessive allele. Thus, in the example above dominant purple flower allele will hide the phenotypic effects of the recessive white flower allele. This is known as the Law of Dominance but it is not a transmission law, dominance has to do with the expression of the genotype and not its transmission. The upper case letters are used to represent dominant alleles whereas the lowercase letters are used to represent recessive alleles.

Mendel's laws of inheritance	
Law	**Definition**
Law of segregation	During gamete formation, the alleles for each gene segregate from each other so that each gamete carries only one allele for each gene.
Law of independent assortment	Genes for different traits can segregate independently during the formation of gametes.
Law of dominance	Some alleles are dominant while others are recessive; an organism with at least one dominant allele will display the effect of the dominant allele.

In the pea plant example above, the capital "P" represents the dominant allele for purple flowers and lowercase "p" represents the recessive allele for white flowers. Both parental plants were true-breeding, and one parental variety had two alleles for purple flowers (*PP*) while the other had two alleles for white flowers (*pp*). As a result of fertilization, the F_1 hybrids each inherited one allele for purple flowers and one for white. All the F_1 hybrids (*Pp*) had purple flowers, because the dominant *P* allele has its full effect in the heterozygote, while the recessive *p* allele has no effect on flower color. For the F_2 plants, the ratio of plants with purple flowers to those with white flowers (3:1) is called the phenotypic ratio. The genotypic ratio, as seen in the Punnett square, is 1 *PP* : 2 *Pp* : 1 *pp*.

Law of Segregation of Genes (the "First Law")

The Law of Segregation states that every individual organism contains two alleles for each trait, and that these alleles segregate (separate) during meiosis such that each gamete contains only one of the alleles. An offspring thus receives a pair of alleles for a trait by inheriting homologous chromosomes from the parent organisms: one allele for each trait from each parent.

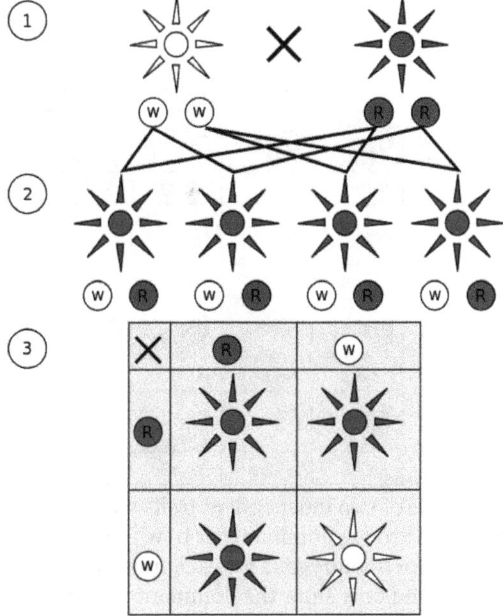

Figure 1 Dominant and recessive phenotypes.
(1) Parental generation.
(2) F_1 generation.
(3) F_2 generation. Dominant (red) and recessive (white) phenotype look alike in the F_1 (first) generation and show a 3:1 ratio in the F_2 (second) generation.

Molecular proof of this principle was subsequently found through observation of meiosis by two scientists independently, the German botanist Oscar Hertwig in 1876, and the Belgian zoologist Edouard Van Beneden in 1883. Paternal and maternal chromosomes get separated in meiosis and the alleles with the traits of a character are segregated into two different gametes. Each parent contributes a single gamete, and thus a single, randomly successful allele copy to their offspring and fertilization.

Law of Independent Assortment (the "Second Law")

The Law of Independent Assortment states that alleles for separate traits are passed independently of one another from parents to offspring. That is, the biological selection of an allele for one trait has nothing to do with the selection of an allele for any other trait. Mendel found support for this law in his dihybrid cross experiments (Fig. 1). In his monohybrid crosses, an idealized 3:1 ratio between dominant and recessive phenotypes resulted. In dihybrid crosses, however, he found a 9:3:3:1 ratios (Fig. 2). This shows that each of the two alleles is inherited independently from the other, with a 3:1 phenotypic ratio for each.

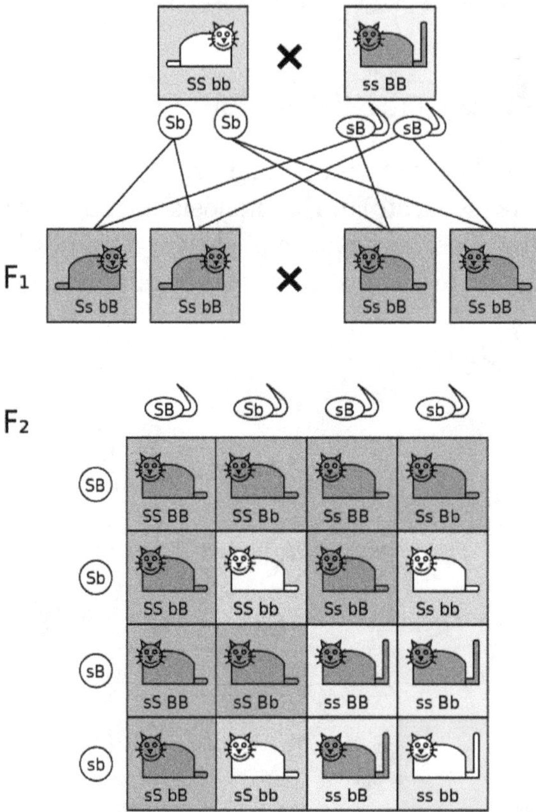

Figure 2 Dihybrid cross. The phenotypes of two independent traits show a 9:3:3:1 ratio in the F_2 generation. In this example, coat color is indicated by B (brown, dominant) or b (white), while tail length is indicated by S (short, dominant) or s (long). When parents are homozygous for each trait (SSbb and ssBB), their children in the F_1 generation are heterozygous at both loci and only show the dominant phenotypes (SsbB). If the children mate with each other, in the F_2 generation all combinations of coat color and tail length occur: 9 are brown/short (purple boxes), 3 are white/short (pink boxes), 3 are brown/long (blue boxes) and 1 is white/long (green box).

Independent assortment occurs in eukaryotic organisms during meiotic prophase I, and produces a gamete with a mixture of the organism's chromosomes. The physical basis of the independent assortment of chromosomes is the random orientation of each bivalent chromosome along the metaphase plate with respect to the other bivalent chromosomes. Along with crossing over, independent assortment increases genetic diversity by producing novel genetic combinations.

There are many violations of independent assortment due to genetic linkage.

Of the 46 chromosomes in a normal diploid human cell, half are maternally derived (from the mother's egg) and half are paternally derived (from the father's sperm). This occurs as sexual reproduction involves the fusion of two haploid gametes (the egg and sperm) to produce a new organism having the full complement of chromosomes. During gametogenesis—the production of new gametes by an adult—the normal complement of 46 chromosomes needs to be halved to 23 to ensure that the resulting haploid gamete can join with another gamete to produce a diploid organism. An error in the number of chromosomes, such as those caused by a diploid gamete joining with a haploid gamete, is termed aneuploidy.

In independent assortment, the chromosomes that result are randomly sorted from all possible maternal and paternal chromosomes. Because zygotes end up with a random mix instead of a pre-defined "set" from either parent, chromosomes are therefore considered assorted independently. As such, the zygote can end up with any combination of paternal or maternal chromosomes. Any of the possible variants of a zygote formed from maternal and paternal chromosomes will occur with equal frequency. For human gametes, with 23 pairs of chromosomes, the number of possibilities is 2^{23} or 8,388,608 possible combinations. The zygote will normally end up with 23 chromosomes pairs, but the origin of any particular chromosome will be randomly selected from paternal or maternal chromosomes. This contributes to the genetic variability of progeny.

Law of Dominance (the "Third Law")

Mendel's Law of Dominance states that recessive alleles will always be masked by dominant alleles. Therefore, a cross between a homozygous dominant and a homozygous recessive will always express the dominant phenotype, while still having a heterozygous genotype. Law of Dominance can be explained easily with the help of a mono hybrid cross experiment:- In a cross between two organisms pure for any pair (or pairs) of contrasting traits (characters), the character that appears in the F1 generation is called "dominant" and the one which is suppressed (not expressed) is called "recessive." Each character is controlled by a pair of dissimilar factors. Only one of the characters expresses. The one which expresses in the F1 generation is called Dominant. It is important to note however, that the law of dominance is significant and true but is not universally applicable.

According to the latest revisions, only two of these rules are considered to be laws. The third one is considered as a basic principle but not a genetic law of Mendel.

Mendelian Trait

A Mendelian trait is one that is controlled by a single locus in an inheritance pattern. In such cases, a mutation in a single gene can cause a disease that is inherited according to Mendel's laws. Examples include sickle-cell anemia, Tay-Sachs disease, cystic fibrosis and xeroderma pigmentosa. A disease controlled by a single gene contrasts with a multi-factorial disease, like arthritis, which is affected by several loci (and the environment) as well as those diseases inherited in a non-Mendelian fashion.

Non-Mendelian Inheritance

Mendel explained inheritance in terms of discrete factors—genes—that are passed along from generation to generation according to the rules of probability. Mendel's laws are valid for all sexually reproducing organisms, including garden peas and human beings. However, Mendel's laws stop short of explaining some patterns of genetic inheritance. For most sexually reproducing organisms, cases where Mendel's laws can strictly account for the patterns of inheritance are relatively rare. Often, the inheritance patterns are more complex.

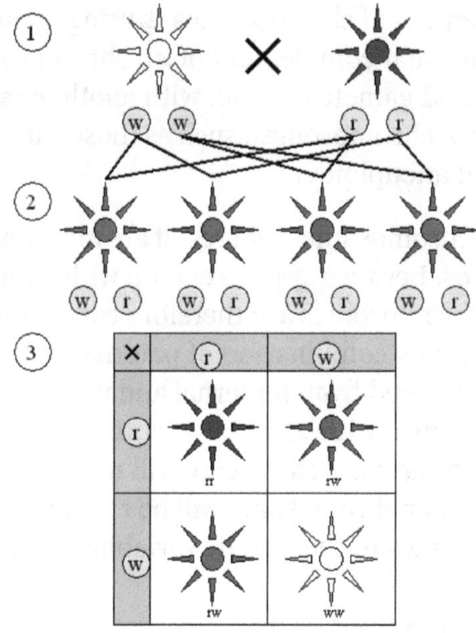

In four o'clock plants, the alleles for red and white flowers show incomplete dominance. As seen in the F_1 generation, heterozygous (*wr*) plants have "pink" flowers—a mix of "red" (rr) and "white" (ww) coloring. The F_2 generation shows a 1:2:1 ratio of red:pink:white

The F_1 offspring of Mendel's pea crosses always looked like one of the two parental varieties. In this situation of "complete dominance," the dominant allele had the same phenotypic effect whether present in one or two copies. But for some characteristics, the F_1 hybrids have an appearance *in between* the phenotypes of the two parental varieties. A cross between two four o'clock (*Mirabilis jalapa*) plants shows this common exception to Mendel's principles. Some alleles are neither dominant nor recessive. The F_1 generation produced by a cross between red-flowered (RR) and white flowered (WW) *Mirabilis jalapa* plants consists of pink-colored flowers (RW). Which allele is dominant in this case? Neither one. This third phenotype results from flowers of the heterzygote having less red pigment than the red homozygotes. Cases in which one allele is not completely dominant over another are called incomplete dominance. In incomplete dominance, the heterozygous phenotype lies somewhere between the two homozygous phenotypes.

A similar situation arises from codominance, in which the phenotypes produced by both alleles are clearly expressed. For example, in certain varieties of chicken, the allele for black feathers is codominant with the allele for white feathers. Heterozygous chickens have a color described as "erminette," speckled with black and white feathers. Unlike the blending of red and white colors in heterozygous four o'clocks, black and white colors appear separately in chickens. Many human genes, including one for a protein that controls cholesterol levels in the blood, show codominance, too. People with the heterozygous form of this gene produce two different forms of the protein, each with a different effect on cholesterol levels.

In Mendelian inheritance, genes have only two alleles, such as *a* and *A*. In nature, such genes exist in several different forms and are therefore said to have multiple alleles. A gene with more than two alleles is said to have multiple alleles. An individual, of course, usually has only two copies of each gene, but many different alleles are often found within a population. One of the best-known examples is coat color in rabbits. A rabbit's coat color is determined by a single gene that has at least four different alleles. The four known alleles display a pattern of simple dominance that can produce four coat colors. Many other genes have multiple alleles, including the human genes for ABO blood type.

Furthermore, many traits are produced by the interaction of several genes. Traits controlled by two or more genes are said to be polygenic traits. *Polygenic* means "many genes." For example, at least three genes are involved in making the reddish-brown pigment in the eyes of fruit flies. Polygenic traits often show a wide range of phenotypes. The broad variety of skin color in humans comes about partly because at least four different genes probably control this trait.

Heredity

Heredity is the genetic information passing for traits from parents to their offspring, either through asexual reproduction or sexual reproduction. This is the process by which an offspring cell or organism acquires or becomes predisposed to the characteristics of its parent cell or organism. Through heredity, variations exhibited by individuals can accumulate and cause some species to evolve through the natural selection of specific phenotype traits. The study of heredity in biology is called genetics, which includes the field of epigenetics.

Overview

In humans, eye color is an example of an inherited characteristic: an individual might inherit the "brown-eye trait" from one of the parents. Inherited traits are controlled by genes and the complete set of genes within an organism's genome is called its genotype.

Heredity of phenotypic traits: Father and son with prominent ears and crowns.

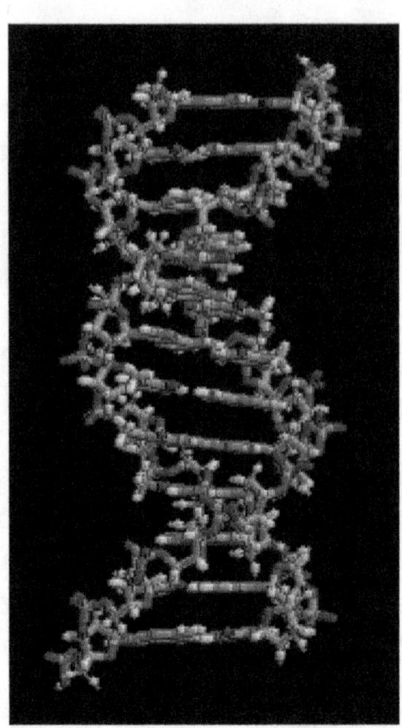

DNA structure. Bases are in the centre, surrounded by phosphate–sugar chains in a double helix.

The complete set of observable traits of the structure and behavior of an organism is called its phenotype. These traits arise from the interaction of its genotype with the environment. As a result, many aspects of an organism's phenotype are not inherited. For example, suntanned skin comes from the interaction between a person's phenotype and sunlight; thus, suntans are not passed on to people's children. However, some people tan more easily than others, due to differences in their genotype: a striking example is people with the inherited trait of albinism, who do not tan at all and are very sensitive to sunburn.

Heritable traits are known to be passed from one generation to the next via DNA, a molecule that encodes genetic information. DNA is a long polymer that incorporates four types of bases, which are interchangeable. The sequence of bases along a particular DNA molecule specifies the genetic information: this is comparable to a sequence of letters spelling out a passage of text. Before a cell divides through mitosis, the DNA is copied, so that each of the resulting two cells will inherit the DNA sequence. A portion of a DNA molecule that specifies a single functional unit is called a gene; different genes have different sequences of bases. Within cells, the long strands of DNA form condensed structures called chromosomes. Organisms inherit genetic material from their parents in the form of homologous chromosomes, containing a unique combination of DNA sequences that code for genes. The specific location of a DNA sequence within a chromosome is known as a locus. If the DNA sequence at a particular locus varies between individuals, the different forms of this sequence are called alleles. DNA sequences can change through mutations, producing new alleles. If a mutation occurs within a gene, the new allele may affect the trait that the gene controls, altering the phenotype of the organism.

However, while this simple correspondence between an allele and a trait works in some cases, most traits are more complex and are controlled by multiple interacting genes within and among organisms. Developmental biologists suggest that complex interactions in genetic networks and communication among cells can lead to heritable variations that may underlie some of the mechanics in developmental plasticity and canalization.

Recent findings have confirmed important examples of heritable changes that cannot be explained by direct agency of the DNA molecule. These phenomena are classed as epigenetic inheritance systems that are causally or independently evolving over genes. Research into modes and mechanisms of epigenetic inheritance is still in its scientific infancy, however, this area of research has attracted much recent activity as it broadens the scope of heritability and evolutionary biology in general. DNA methylation marking chromatin, self-sustaining metabolic loops, gene silencing by RNA interference, and the three dimensional conformation of proteins (such as prions) are areas where epigenetic inheritance systems have been discovered at the organismic level. Heritability may also occur at even larger scales. For example, ecological inheritance through the process of niche construction is defined by the regular and repeated activities of organisms in their environment. This generates a legacy of effect that modifies and feeds back into the selection regime of subsequent generations. Descendants inherit genes plus environmental characteristics generated by the ecological actions of ancestors. Other examples of heritability in evolution that are not under the direct control of genes include the inheritance of cultural traits, group heritability, and symbiogenesis. These examples of heritability that operate above the gene are covered broadly under the title of multilevel or hierarchical selection, which has been a subject of intense debate in the history of evolutionary science.

Relation to Theory of Evolution

When Charles Darwin proposed his theory of evolution in 1859, one of its major problems was the lack of an underlying mechanism for heredity. Darwin believed in a mix of blending inheritance and the inheritance of acquired traits (pangenesis). Blending inheritance would lead to uniformity across populations in only a few generations and then would remove variation from a population on which natural selection could act. This led to Darwin adopting some Lamarckian ideas in later editions of *On the Origin of Species* and his later biological works. Darwin's primary approach to heredity was to outline how it appeared to work (noticing that traits that were not expressed explicitly in the parent at the time of reproduction could be inherited, that certain traits could be sex-linked, etc.) rather than suggesting mechanisms.

Darwin's initial model of heredity was adopted by, and then heavily modified by, his cousin Francis Galton, who laid the framework for the biometric school of heredity. Galton found no evidence to support the aspects of Darwin's pangenesis model, which relied on acquired traits.

The inheritance of acquired traits was shown to have little basis in the 1880s when August Weismann cut the tails off many generations of mice and found that their offspring continued to develop tails.

History

Scientists in Antiquity had a variety of ideas about heredity: Theophrastus proposed that male flowers caused female flowers to ripen; Hippocrates speculated that "seeds" were produced by various body parts and transmitted to offspring at the time of conception; and Aristotle thought that male and female semen mixed at conception. Aeschylus, in 458 BC, proposed the male as the parent, with the female as a "nurse for the young life sown within her".

Ancient understandings of heredity transitioned to two debated doctrines in the 18th century. The Doctrine of Epigenesis and the Doctrine of Preformation were two distinct views of the understanding of heredity. The Doctrine of Epigenesis, originated by Aristotle, claimed that an embryo continually develops. The modifications of the parent's traits are passed off to an embryo during its lifetime. The foundation of this doctrine was based on the theory of inheritance of acquired traits. In direct opposition, the Doctrine of Preformation claimed that "like generates like" where the germ would evolve to yield offspring similar to the parents. The Preformationist view believed procreation was an act of revealing what had been created long before. However, this was disputed by the creation of the cell theory in the 19th century, where the fundamental unit of life is the cell, and not some preformed parts of an organism. Various hereditary mechanisms, including blending inheritance were also envisaged without being properly tested or quantified, and were later disputed. Nevertheless, people were able to develop domestic breeds of animals as well as crops through artificial selection. The inheritance of acquired traits also formed a part of early Lamarckian ideas on evolution.

During the 18th century, Dutch microscopist Antonie van Leeuwenhoek (1632–1723) discovered "animalcules" in the sperm of humans and other animals. Some scientists speculated they saw a "little man" (homunculus) inside each sperm. These scientists formed a school of thought known as the "spermists". They contended the only contributions of the female to the next generation were the womb in which the homunculus grew, and prenatal influences of the womb. An opposing school of thought, the ovists, believed that the future human was in the egg, and that sperm merely stimulated the growth of the egg. Ovists thought women carried eggs containing boy and girl children, and that the gender of the offspring was determined well before conception.

Gregor Mendel: Father of Genetics

The idea of particulate inheritance of genes can be attributed to the Moravian monk Gregor Mendel who published his work on pea plants in 1865. However, his work was not widely known and was rediscovered in 1901. It was initially assumed the Mendelian inheritance only accounted for large (qualitative) differences, such as those seen by Mendel in his pea plants—and the idea of additive effect of (quantitative) genes was not realised until R. A. Fisher's (1918) paper, "The Correlation Between Relatives on the Supposition of Mendelian Inheritance" Mendel's overall contribution gave scientists a useful overview that traits were inheritable. His pea plant demonstration became the foundation of the study of Mendelian Traits. These traits can be traced on a single locus.

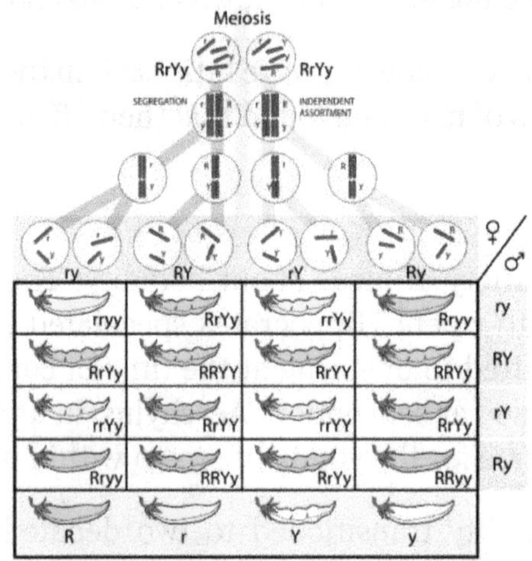

Table showing how the genes exchange according to segregation or independent assortment during meiosis and how this translates into Mendel's laws

Modern Development of Genetics and Heredity

In the 1930s, work by Fisher and others resulted in a combination of Mendelian and biometric schools into the modern evolutionary synthesis. The modern synthesis bridged the gap between experimental geneticists and naturalists; and between both and palaeontologists, stating that:

1. All evolutionary phenomena can be explained in a way consistent with known genetic mechanisms and the observational evidence of naturalists.
2. Evolution is gradual: small genetic changes, recombination ordered by natural selection. Discontinuities amongst species (or other taxa) are explained as originating gradually through geographical separation and extinction (not saltation).
3. Selection is overwhelmingly the main mechanism of change; even slight advantages are important when continued. The object of selection is the phenotype in its surrounding environment. The role of genetic drift is equivocal; though strongly supported initially by Dobzhansky, it was downgraded later as results from ecological genetics were obtained.
4. The primacy of population thinking: the genetic diversity carried in natural populations is a key factor in evolution. The strength of natural selection in the wild was greater than expected; the effect of ecological factors such as niche occupation and the significance of barriers to gene flow are all important.

The idea that speciation occurs after populations are reproductively isolated has been much debated. In plants, polyploidy must be included in any view of speciation. Formulations such as 'evolution consists primarily of changes in the frequencies of alleles between one generation and another' were proposed rather later. The traditional view is that developmental biology ('evo-devo') played little part in the synthesis, but an account of Gavin de Beer's work by Stephen Jay Gould suggests he may be an exception.

Almost all aspects of the synthesis have been challenged at times, with varying degrees of success. There is no doubt, however, that the synthesis was a great landmark in evolutionary biology. It cleared up many confusions, and was directly responsible for stimulating a great deal of research in the post-World War II era.

Trofim Lysenko however caused a backlash of what is now called Lysenkoism in the Soviet Union when he emphasised Lamarckian ideas on the inheritance of acquired traits. This movement affected agricultural research and led to food shortages in the 1960s and seriously affected the USSR.

Common Genetic Disorders

- Down syndrome
- sickle cell disease
- Phenylketonuria (PKU)
- Haemophilia

Types

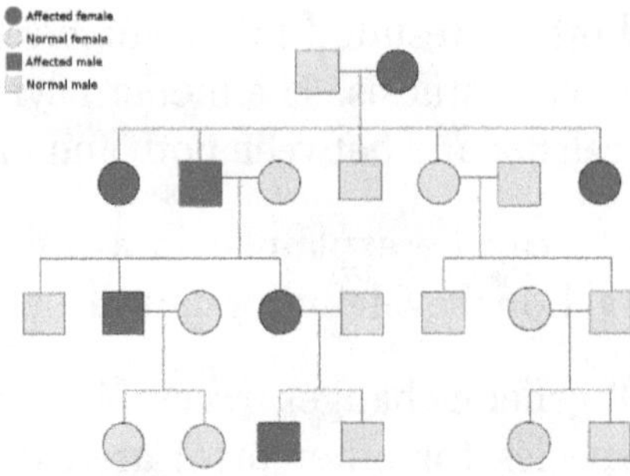

An example pedigree chart of an autosomal dominant disorder.

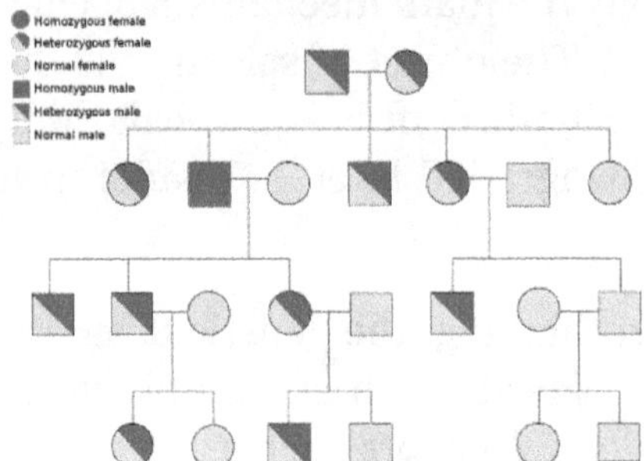

An example pedigree chart of an autosomal recessive disorder.

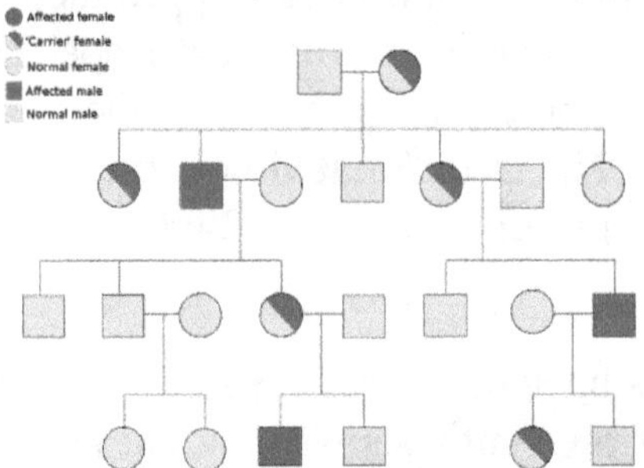

An example pedigree chart of a sex-linked disorder (the gene is on the X chromosome)

Dominant and Recessive Alleles

An allele is said to be dominant if it is always expressed in the appearance of an organism (phenotype) provided that at least one copy of it is present. For example, in peas the allele for green pods, *G*, is dominant to that for yellow pods, *g*. Thus pea plants with the pair of alleles *either GG* (homozygote) ***or*** *Gg* (heterozygote) will have green pods. The allele for yellow pods is recessive. The effects of this allele are only seen when it is present in both chromosomes, *gg* (homozygote).

The description of a mode of biological inheritance consists of three main categories:

1. Number of involved loci
    - Monogenetic (also called "simple") – one locus

- Oligogenetic – few loci
- Polygenetic – many loci

2. Involved chromosomes
    - Autosomal – loci are not situated on a sex chromosome
    - Gonosomal – loci are situated on a sex chromosome
        - X-chromosomal – loci are situated on the X-chromosome (the more common case)
        - Y-chromosomal – loci are situated on the Y-chromosome
    - Mitochondrial – loci are situated on the mitochondrial DNA
3. Correlation genotype–phenotype
    - Dominant
    - Intermediate (also called “codominant”)
    - Recessive
    - Overdominant
    - Underdominant

These three categories are part of every exact description of a mode of inheritance in the above order. In addition, more specifications may be added as follows:

4. Coincidental and environmental interactions
    - Penetrance
        - Complete
        - Incomplete (percentual number)
    - Expressivity
        - Invariable
        - Variable
    - Heritability (in polygenetic and sometimes also in oligogenetic modes of inheritance)
    - Maternal or paternal imprinting phenomena
5. Sex-linked interactions
    - Sex-linked inheritance (gonosomal loci)

- Sex-limited phenotype expression (e.g., cryptorchism)
- Inheritance through the maternal line (in case of mitochondrial DNA loci)
- Inheritance through the paternal line (in case of Y-chromosomal loci)

6. Locus–locus interactions
    - Epistasis with other loci (e.g., overdominance)
    - Gene coupling with other loci
    - Homozygotous lethal factors
    - Semi-lethal factors

Determination and description of a mode of inheritance is achieved primarily through statistical analysis of pedigree data. In case the involved loci are known, methods of molecular genetics can also be employed.

References

- Alberts B, Johnson A, Lewis J, Raff M, Roberts K, Walter P (2002). Molecular Biology of the Cell (Fourth ed.). New York: Garland Science. ISBN 978-0-8153-3218-3.
- Henig, Robin Marantz (2000). The Monk in the Garden: The Lost and Found Genius of Gregor Mendel, the Father of Genetics. Boston: Houghton Mifflin. pp. 1–9. ISBN 978-0395-97765-1.
- Judson, Horace (1979). The Eighth Day of Creation: Makers of the Revolution in Biology. Cold Spring Harbor Laboratory Press. pp. 51–169. ISBN 0-87969-477-7.
- Huxley, Julian (1942). Evolution: the modern synthesis (Definitive ed.). Cambridge, Mass.: MIT Press. ISBN 978-0262513661.
- Williams, George C. (2001). Adaptation and Natural Selection a Critique of Some Current Evolutionary Thought. ([Online-Ausg.]. ed.). Princeton: Princeton University Press. ISBN 9781400820108.
- Domingo, Esteban (2001). "RNA Virus Genomes". ELS. doi:10.1002/9780470015902.a0001488.pub2. ISBN 0470016175.
- Iverson, Cheryl, et al. (eds) (2007). "15.6.1 Nucleic Acids and Amino Acids". AMA Manual of Style (10th ed.). Oxford, Oxfordshire: Oxford University Press. ISBN 978-0-19-517633-9.
- Iverson, Cheryl, et al. (eds) (2007). "15.6.2 Human Gene Nomenclature". AMA Manual of Style (10th ed.). Oxford, Oxfordshire: Oxford University Press. ISBN 978-0-19-517633-9.
- Ridley, M. (2006), Genome: the autobiography of a species in 23 chapters (PDF), New York: Harper Perennial, ISBN 0-06-019497-9
- Griffiths JF; Gelbart WM; Lewontin RC; Wessler SR; Suzuki DT; Miller JH (2005). Introduction to Genetic Analysis. New York: W.H. Freeman and Co. pp. 34–40, 473–476, 626–629. ISBN 0-7167-4939-4.
- Koonin, Eugene V. (2011-08-31). The Logic of Chance: The Nature and Origin of Biological Evolution. FT Press. ISBN 9780132542494.
- Mankertz P (2008). "Molecular Biology of Porcine Circoviruses". Animal Viruses: Molecular Biology. Caister Academic Press. ISBN 978-1-904455-22-6.
- Stojanovic, edited by Nikola (2007). Computational genomics : current methods. Wymondham: Horizon Bioscience. ISBN 1-904933-30-0.

5

An Integrated Study of Artificial Insemination

The process of reproducing by means other than sexual intercourse is referred to as artificial insemination. It's a deliberate introduction of sperm into the female's uterus. As compared to the natural way of reproducing, artificial insemination is expensive and involves professional help. The application of artificial insemination in the current scenario has been thoroughly discussed in this chapter.

Artificial Insemination

Artificial insemination (AI) is the deliberate introduction of sperm into a female's uterus or cervix for the purpose of achieving a pregnancy through in vivo fertilization by means other than sexual intercourse. It is a fertility treatment for humans, and is a common practice in animal breeding, including dairy cattle and pigs.

Artificial insemination may employ assisted reproductive technology, sperm donation and animal husbandry techniques. Artificial insemination techniques available include intracervical insemination and intrauterine insemination. The primary beneficiaries of artificial insemination are heterosexual couples suffering from male infertility, lesbian couples and single women. Intracervical insemination (ICI) is the easiest and most common insemination technique and can be used in the home for self-insemination without medical practitioner assistance. Compared to natural insemination (i.e., insemination by sexual intercourse), artificial insemination can be more expensive and more invasive, and may require professional assistance.

There are laws in some countries which restrict and regulate who can donate sperms and who is able to receive artificial insemination, and the consequences of such insemination. Subject to any regulations restricting who can obtain donor sperms, donor sperms are available to all women who, for whatever reason, want or need them. Some women living in a jurisdiction which does not permit artificial insemination in the circumstance in which she finds herself may travel to another jurisdiction which permits it.

In Humans

General

Before artificial insemination is turned to as the solution to impregnate a woman, doctors will require an examination of both the male and female involved in order to remove any and all physical hindrances that are preventing them from naturally achieving a pregnancy. The couple is also given a fertility test to determine the motility, number, and viability of the male's sperm and the success of the female's ovulation. From these tests, the doctor may or may not recommend a form of artificial insemination.

The sperm used in artificial insemination may be provided by either the woman's husband or partner (partner sperm) or by a known or anonymous sperm donation (donor sperm). Though there may be legal, religious and cultural differences in these and other characterizations, the manner in which the sperm is actually used in AI would be the same, If the procedure is successful, the woman will conceive and carry a baby to term in the normal manner. A pregnancy resulting from artificial insemination will be no different from a pregnancy achieved by sexual intercourse. In all cases, the woman would be the biological mother of any child produced by AI, and the male whose sperm is used would be the biological father.

There are multiple methods used to obtain the semen necessary for artificial insemination. Some methods require only men, while others require a combination of a male and female. Those that require only men to obtain semen are masturbation, a rectal massage, involuntary pollution (the collecting of nocturnal emission), or the aspiration of sperm by means of a puncture of the testicle and epididymus. Methods of collecting semen that involve a combination of a male and female include interrupted intercourse, or the withdrawal just before ejaculation, or the post coital aspiration of the semen from the vagina.

There are a number of reasons why a woman would use artificial insemination to achieve pregnancy. For example, a woman's immune system may be rejecting her partner's sperm as invading molecules. Women who have issues with the cervix, such as cervical scarring, cervical blockage from endometriosis, or thick cervical mucus may also benefit from artificial insemination since the sperm must pass through the cervix to result in fertilization.

Donor sperm may be used if the woman's male partner is impotent, when a lesbian couple wish to have a biological child, a couple where one person is transgender and no longer has gonads, or by a single woman who does not have or want a male partner. Partner sperm may be used when a male partner's physical limitation impedes his ability to impregnate her by sexual intercourse, or the partner's sperm had been frozen in anticipation of some medical procedure or if the partner has since died.

Preparations

Timing is critical, as the window and opportunity for fertilization is little more than twelve hours from the release of the ovum. To increase the chance of success, the woman's menstrual cycle is closely observed, often using ovulation kits, ultrasounds or blood tests, such as basal body temperature tests over, noting the color and texture of the vaginal mucus, and the softness of the nose of her cervix. To improve the success rate of AI, drugs to create a stimulated cycle may be used, but the use of such drugs also results in an increased chance of a multiple birth.

Sperm can be provided fresh or washed. The washing of sperm increases the chances of fertilization. Pre- and post-concentration of motile sperm is counted. Sperm from a sperm bank will be frozen and quarantined for a period and the donor will be tested before and after production of the sample to ensure that he does not carry a transmissible disease. For fresh shipping, a semen extender is used.

If sperm is provided by a private donor, either directly or through a sperm agency, it is usually supplied fresh, not frozen, and it will not be quarantined. Donor sperm provided in this way may be given directly to the recipient woman or her partner, or it may be transported in specially insulated containers. Some donors have their own freezing apparatus to freeze and store their sperm.

Techniques

Semen used in insemination would be used either fresh, raw or frozen. Where donor sperm is supplied by a sperm bank, it will always be quarantined and frozen and will need to be thawed before use. When an ovum is released, semen is introduced into the woman's vagina, uterus or cervix, depending on the method being used.

Intracervical Insemination

Intracervical insemination (ICI) involves injection of unwashed or raw semen into the cervix with a needleless syringe. Where fresh semen is used this must be allowed to liquefy before inserting it into the syringe, or alternatively, the syringe may be back-loaded. After the syringe has been filled with semen, enclosed air is removed by gently pressing the plunger forward before inserting the syringe into the vagina. Care is optimal when inserting the syringe, so that the tip is as close to the entrance to the cervix as possible. A vaginal speculum may be used to hold open the vagina so that the cervix may be observed and the syringe inserted more accurately through the open speculum. The plunger is then pushed forward and the semen in the syringe is emptied deep into the vagina. The syringe (and speculum if used) may be left in place for several minutes and the woman is advised to lie still for about half-an-hour to improve the success rate.

ICI is the easiest and most common insemination technique. The process closely replicates natural insemination, with fresh semen being directly deposited onto the neck of the cervix. It is the simplest artificial insemination method and unwashed or raw semen is normally used, but frozen semen (which has been thawed) may also be used. It is commonly used in home, self-insemination and practitioner insemination procedures, and for insemination where semen is provided by private donors. When performed at home without the presence of a professional this procedure is sometimes referred to as intravaginal insemination (IVI).

A conception cap, which is a form of conception device may be inserted into the vagina following insemination and may be left in place near to the entrance to the cervix for several hours. Using this method, a woman may go about her usual activities while the cervical cap holds the semen in the vagina. One advantage with the conception device is that fresh, non-liquefied semen may be used. Other methods may be used to insert semen into the vagina notably involving different uses of a conception cap. This may, for example, be inserted filled with sperm which does not have to be liquefied. The male may therefore ejaculate straight into the cap. Alternatively, a specially designed conception cap with a tube attached may be inserted empty into the vagina after which liquefied semen is poured into the tube. These methods are designed to ensure that donor or partner semen is inseminated as close as possible to the cervix and that it is kept in place there to increase the chances of conception.

Intrauterine Insemination

Intrauterine insemination (IUI) involves injection of washed sperm into the uterus with a catheter. If unwashed semen is used, it may elicit uterine cramping, expelling the semen and causing pain, due to content of prostaglandins. (Prostaglandins are also the compounds responsible for causing the myometrium to contract and expel the menses from the uterus, during menstruation.) Resting on the table for fifteen minutes after an IUI is optimal for the woman to increase the pregnancy rate.

Unlike ICI, intrauterine insemination normally requires a medical practitioner to perform the procedure. A female under 30 years of age has optimal chances with IUI; for the man, a TMS of more than 5 million per ml is optimal. In practice, donor sperm will satisfy these criteria. A promising cycle is one that offers two follicles measuring more than 16 mm, and estrogen of more than 500 pg/mL on the day of hCG administration. A short period of ejaculatory abstinence before intrauterine insemination is associated with higher pregnancy rates. However, GnRH agonist administration at the time of implantation does not improve pregnancy outcome in intrauterine insemination cycles according to a randomized controlled trial.

IUI is a more efficient method of artificial insemination. Sperm is occasionally inserted twice within a 'treatment cycle'. A double intrauterine insemination theoretically increases pregnancy rates by decreasing the risk of missing the fertile window during ovulation. However, a randomized trial of insemination after ovarian hyperstimulation found no difference in live birth rate between single and double intrauterine insemination.

IUI can be used in conjunction with controlled ovarian hyperstimulation (COH). Still, advanced maternal age causes decreased success rates; women aged 38–39 years appear to have reasonable success during the first two cycles of ovarian hyperstimulation and IUI. However, for women aged over 40 years, there appears to be no benefit after a single cycle of COH/IUI. Medical experts therefore recommend considering in vitro fertilization after one failed COH/IUI cycle for women aged over 40 years.

Intrauterine Tuboperitoneal Insemination

Intrauterine tuboperitoneal insemination (IUTPI) involves injection of washed sperm into both the uterus and fallopian tubes. The cervix is then clamped to prevent leakage to the vagina, best achieved with a specially designed double nut bivalve (DNB) speculum. The sperm is mixed to create a volume of 10 ml, sufficient to fill the uterine cavity, pass through the interstitial part of the tubes and the ampulla, finally reaching the peritoneal cavity and the Pouch of Douglas where it would be mixed with the peritoneal and follicular fluid. IUTPI can be useful in unexplained infertility, mild or moderate male infertility, and mild or moderate endometriosis. In non-tubal sub fertility, fallopian tube sperm perfusion may be the preferred technique over intrauterine insemination.

Intratubal Insemination

Intratubal insemination (ITI) involves injection of washed sperm into the fallopian tube, although this procedure is no longer generally regarded as having any beneficial effect compared with IUI. ITI however, should not be confused with gamete intrafallopian transfer, where both eggs and sperm are mixed outside the woman's body and then immediately inserted into the fallopian tube where fertilization takes place.

Pregnancy Rate

The pregnancy or success rates for artificial insemination are 10 to 15% per menstrual cycle using ICI, and and 15–20% per cycle for IUI. In IUI, about 60 to 70% have achieved pregnancy after 6 cycles.

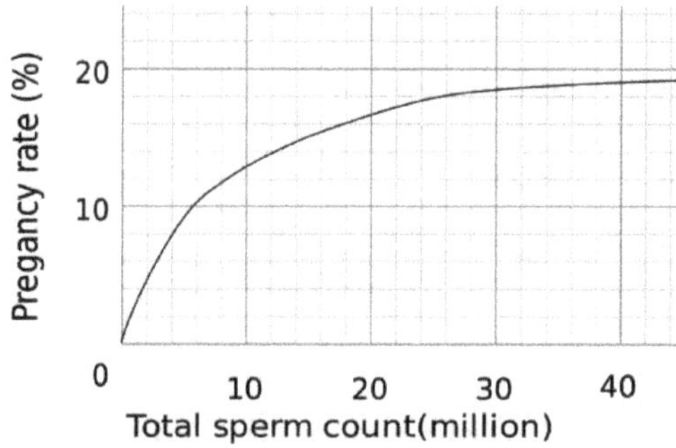

Approximate pregnancy rate as a function of total sperm count (may be twice as large as total motile sperm count). Values are for intrauterine insemination. (Old data, rates are likely higher today)

However, these pregnancy rates may be very misleading, since many factors, including the age and health of the recipient, have to be included to give a meaningful answer, e.g. definition of success and calculation of the total population. For couples with unexplained infertility, unstimulated IUI is no more effective than natural means of conception.

The pregnancy rate also depends on the total sperm count, or, more specifically, the total motile sperm count (TMSC), used in a cycle. The success rate increases with increasing TMSC, but only up to a certain count, when other factors become limiting to success. The summed pregnancy rate of two cycles using a TMSC of 5 million (may be a TSC of ~10 million on graph) in each cycle is substantially higher than one single cycle using a TMSC of 10 million. However, although more cost-efficient, using a lower TMSC also increases the average time taken to achieve pregnant. Women whose age is becoming a major factor in fertility may not want to spend that extra time.

Samples Per Child

The number of samples (ejaculates) required to give rise to a child varies substantially from person to person, as well as from clinic to clinic. However, the following equations generalize the main factors involved:

For intracervical insemination:

$$N = \frac{V_s \times c \times r_s}{n_r}$$

- N is how many children a single sample can give rise to.
- V_s is the volume of a sample (ejaculate), usually between 1.0 mL and 6.5 mL
- c is the concentration of motile sperm in a sample *after freezing and thawing*, approximately 5–20 million per ml but varies substantially
- r_s is the pregnancy rate per cycle, between 10% to 35%

- n_r is the total motile sperm count recommended for vaginal insemination (VI) or intra-cervical insemination (ICI), approximately 20 million pr. ml.

The pregnancy rate increases with increasing number of motile sperm used, but only up to a certain degree, when other factors become limiting instead.

Derivation of the equation (click at right to view)

In the simplest form, the equation reads:

$$N = \frac{n_s}{n_c} \times r_s$$

N is how many children a single sample can give rise to

n_s is the number of vials produced per sample

n_c is the number of vials used in a cycle

r_s is the pregnancy rate per cycle

n_s can be further split into:

$$n_s = \frac{V_s}{V_v}$$

n_s is the number of vials produced per sample

V_s is the volume of a sample

V_v is the volume of the vials used

n_c may be split into:

$$n_c = \frac{n_r}{n_s}$$

n_c is the number of vials used in a cycle

n_r is the number of motile sperm recommended for use in a cycle

n_s is the number of motile sperm in a vial

n_s may be split into:

$$n_s = V_v \times c$$

n_s is the number of motile sperm in a vial

V_v is the volume of the vials used

c is the concentration of motile sperm in a sample

Thus, the factors can be presented as follows:

$$N = \frac{V_s \times c \times r_s}{n_r} \times \frac{V_v}{V_v}$$

N is how many children a single sample can help giving rise to

V_s is the volume of a sample

c is the concentration of motile sperm in a sample

r_s is the pregnancy rate per cycle

n_r is the number of motile sperm recommended for use in a cycles

V_v is the volume of the vials used (its value doesn't affect N and may be eliminated. In short, the smaller the vials, the more vials are used)

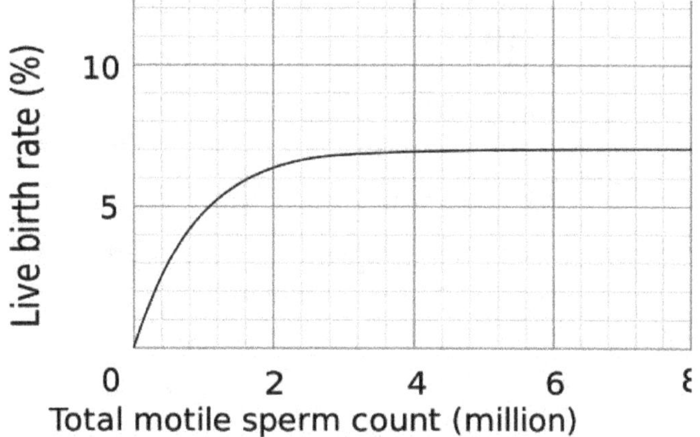

Approximate live birth rate (r_s) among infertile couples as a function of total motile sperm count (n_r). Values are for intrauterine insemination.

With these numbers, one sample would on average help giving rise to 0.1–0.6 children, that is, it actually takes on average 2–5 samples to make a child.

For intrauterine insemination, a *centrifugation fraction* (f_c) may be added to the equation:

f_c is the fraction of the volume that remains after centrifugation of the sample, which may be about half (0.5) to a third (0.33).

$$N = \frac{V_s \times f_c \times c \times r_s}{n_r}$$

On the other hand, only 5 million motile sperm may be needed per cycle with IUI (n_r=5 million)

Thus, only 1–3 samples may be needed for a child if used for IUI.

History

The first reported case of artificial insemination by donor occurred in 1884: a Philadelphia professor of medicine took sperm from his "best looking" student to inseminate an anesthetized woman. The woman was not informed about the procedure, unlike her infertile husband. The case was reported 25 years later in a medical journal. The sperm bank was developed in Iowa starting in the 1920s in research conducted by University of Iowa medical school researchers Jerome Sherman and Raymond Bunge.

In the 1980s, direct intraperitoneal insemination (DIPI) was occasionally used, where doctors injected sperm into the lower abdomen through a surgical hole or incision, with the intention of letting them find the oocyte at the ovary or after entering the genital tract through the ostium of the fallopian tube.

Social Implications

One of the key issues arising from the rise of dependency on assisted reproductive technology (ARTs) is the pressure placed on couples to conceive; 'where children are highly desired, parenthood is culturally mandatory, and childlessness socially unacceptable'.

The medicalization of infertility creates a framework in which individuals are encouraged to think of infertility quite negatively. In many cultures, especially those with large Muslim populations, donor insemination is religiously and culturally prohibited, often meaning that less accessible "high tech" and expensive ARTs, like IVF, are the only solution.

An over-reliance on reproductive technologies in dealing with infertility prevents many – especially, for example, in the "infertility belt" of central and southern Africa – from dealing with many of the key causes of infertility treatable by artificial insemination techniques; namely preventable infections, dietary and lifestyle influences.

Morality in Catholic Church

Within Christianity there are many differing opinions on the morality of the procedures taken for artificial insemination. According to some Catholic literature, "To achieve union but not children by means of contraceptives and to achieve children but not union by means of artificial insemination are both equally wrong." Heterosexual intercourse is viewed by some in the Catholic community as a sacramental act meant to be experienced only by married couples; it is viewed as a physical representation of the spiritual unity of marriage between a husband and wife. Artificial insemination "dissociates the sexual act from the procreative act. The act which brings the child into existence is no longer an act by two persons giving themselves to one another, but one that 'entrusts the life and identity of the embryo into the power of doctors and biologists and establishes the domination of technology over the origin and destiny of the human person. Such a relationship of domination is, in itself, contrary to the dignity and equality that must be common to parents and children'".

Legal Restrictions

Some countries restrict artificial insemination in a variety of ways. For example, some countries do not permit AI for single women, and some Muslim countries do not permit the use of donor sperm.

As of May 2013, the following European countries permit medically assisted AI for single women:

- Armenia
- Belarus
- Belgium
- Bulgaria
- Cyprus
- Denmark
- Estonia
- Finland
- Germany
- Greece
- Hungary
- Iceland
- Ireland
- Latvia
- Macedonia
- Moldova
- Montenegro
- Netherlands
- Romania
- Russia
- Spain
- Ukraine
- United Kingdom

In Livestock and Pets

The first viviparous to be artificially fertilized, was a dog. The experiment was conducted with success by the Italian Lazzaro Spallanzani in 1780. An important pioneer in artificial insemination was also the Russian Ilya Ivanov since 1899. In 1935 Suffolk sheep diluted semen was sent from Cambridge by plane to Krakoiv, Poland, and international research joint (Prawochenki from Poland, Milovanoff from URSS, Hammond from Cambridge, Walton from Scotland, and Thomasset from Uruguay).

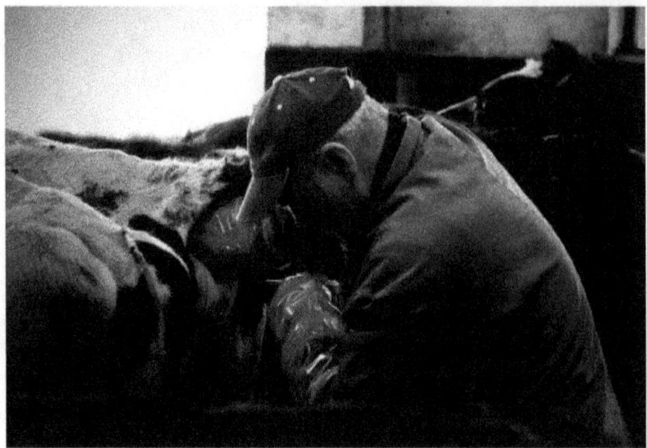

A man performing artificial insemination of a cow.

A breeding mount with built-in artificial vagina used in semen collection from horses for use in artificial insemination

Artificial insemination is used in many non-human animals, including sheep, horses, cattle, pigs, dogs, pedigree animals generally, zoo animals, turkeys and even honeybees. It may be used for many reasons, including to allow a male to inseminate a much larger number of females, to allow use of genetic material from males separated by distance or time, to overcome physical breeding difficulties, to control the paternity of offspring, to synchronise births, to avoid injury incurred during natural mating, and to avoid the need to keep a male at all (such as for small numbers of females or in species whose fertile males may be difficult to manage).

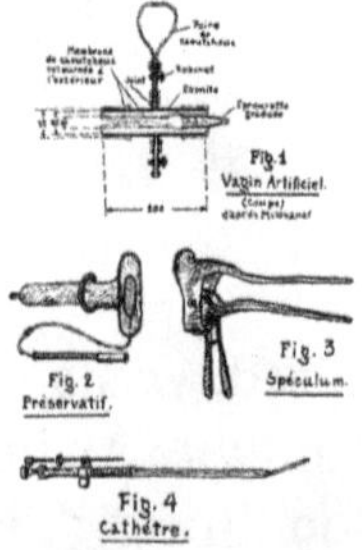

IA tools brought from the USSR by Luis Thomasset in 1935 to work at Cambridge Laboratories and South America.

Semen is collected, extended, then cooled or frozen. It can be used on site or shipped to the female's location. If frozen, the small plastic tube holding the semen is referred to as a *straw*. To allow the sperm to remain viable during the time before and after it is frozen, the semen is mixed with a solution containing glycerol or other cryoprotectants. An *extender* is a solution that allows the semen from a donor to impregnate more females by making insemination possible with fewer sperm. Antibiotics, such as streptomycin, are sometimes added to the sperm to control some bacterial venereal diseases. Before the actual insemination, estrus may be induced through the use of progestogen and another hormone (usually PMSG or Prostaglandin F2α).

Artificial insemination of farm animals is very common in today's agriculture industry in the developed world, especially for breeding dairy cattle (75% of all inseminations). Swine are also bred using this method (up to 85% of all inseminations). It provides an economical means for a livestock breeder to improve their herds utilizing males having very desirable traits.

Although common with cattle and swine, AI is not as widely practised in the breeding of horses. A small number of equine associations in North America accept only horses that have been conceived by "natural cover" or "natural service" – the actual physical mating of a mare to a stallion – the Jockey Club being the most notable of these, as no AI is allowed in Thoroughbred breeding. Other registries such as the AQHA and warmblood registries allow registration of foals created through AI, and the process is widely used allowing the breeding of mares to stallions not resident at the same facility – or even in the same country – through the use of transported frozen or cooled semen.

In modern species conservation, semen collection and artificial insemination is used also in birds. In 2013 scientist of the Justus-Liebig-University of Giessen, Germany, from the working group of Michael Lierz, Clinic for birds, reptiles, amphibians and fish, developed a novel technique for semen collection and artificial insemination in parrots producing the world's first macaw by assisted reproduction (Lierz et al., 2013).

Modern artificial insemination was pioneered by John O. Almquist of the Pennsylvania State University. His improvement of breeding efficiency by the use of antibiotics (first proven with penicillin in 1946) to control bacterial growth, decreasing embrionic mortality and increase fertility, and various new techniques for processing, freezing and thawing of frozen semen significantly enhanced the practical utilization of AI in the livestock industry, and earned him the 1981 Wolf Foundation Prize in Agriculture. Many techniques developed by him have since been applied to other species, including that of the human male.

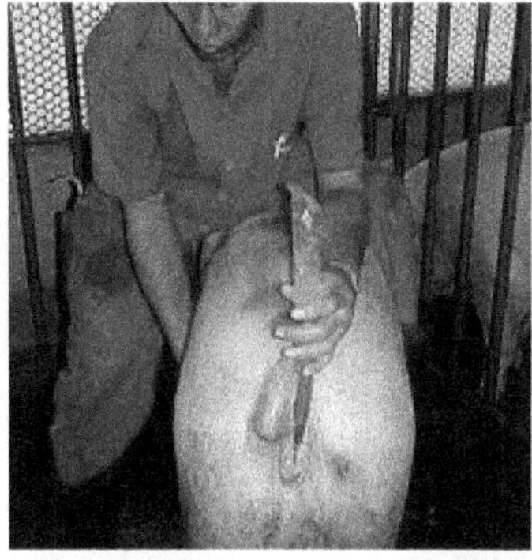

Artificial insemination in pig

Intracytoplasmic Sperm Injection

Intracytoplasmic sperm injection is an *in vitro* fertilization procedure in which a single sperm is injected directly into an egg. Defective sperm function remains the single most important cause of human infertility. Although certain severe forms of male infertility have a genetic origin, others may be the result of environmental factors. During the past decade, ICSI has been applied increasingly around the world to alleviate problems of severe male infertility in human patients who either could not be assisted by conventional IVF procedures or could not be accepted for IVF because too few motile and morphologically normal sperm were present in the ejaculate of the male partner.

Indications

This procedure is most commonly used to overcome male infertility problems, although it may also be used where eggs cannot easily be penetrated by sperm, and occasionally in addition to sperm donation.

It can be used in teratozoospermia, because once the egg is fertilized, abnormal sperm morphology does not appear to influence blastocyst development or blastocyst morphology. Even with severe teratozoospermia, microscopy can still detect the few sperm cells that have a "normal" morphology, allowing for optimal success rate.

History

The first child born from a gamete micromanipulation was a child in Singapore born in April 1989.

The technique was developed by Gianpiero Palermo at the Vrije Universiteit Brussel, in the Center for Reproductive Medicine headed by Paul Devroey and Andre Van Steirteghem.

The procedure itself was first performed in 1987, though it only went to the pronuclear stage. The first activated embryo by ICSI was produced in 1990, but the first successful birth by ICSI took place on January 14, 1992 after an April 1991 conception.

Procedure

ICSI is generally performed following an in vitro fertilization procedure to extract one to several oocytes from a woman.

The procedure is done under a microscope using multiple micromanipulation devices (micromanipulator, microinjectors and micropipettes). A holding pipette stabilizes the mature oocyte with gentle suction applied by a microinjector. From the opposite side a thin, hollow glass micropipette is used to collect a single sperm, having immobilised it by cutting its tail with the point of the micropipette. The oocyte is pierced through the oolemma and directed to the inner part of the oocyte (cytoplasm). The sperm is then released into the oocyte. The pictured oocyte has an extruded polar body at about 12 o'clock indicating its maturity. The polar body is positioned at the 12 or 6 o'clock position, to ensure that the inserted micropipette does not disrupt the spindle inside the egg. After the procedure, the oocyte will be placed into cell culture and checked on the following day for signs of fertilization.

In contrast, in natural fertilization sperm compete and when the first sperm penetrates the oolemma, the oolemma hardens to block the entry of any other sperm. Concern has been raised that in ICSI this sperm selection process is bypassed and the sperm is selected by the embryologist without any specific testing. However, in mid-2006 the FDA cleared a device that allows embryologists to select mature sperm for ICSI based on sperm binding to hyaluronan, the main constituent of the gel layer (cumulus oophorus) surrounding the oocyte. The device provides microscopic droplets of hyaluronan hydrogel attached to the culture dish. The embryologist places the prepared sperm on the microdot, selects and captures sperm that bind to the dot. Basic research on the maturation of sperm shows that hyaluronan-binding sperm are more mature and show fewer DNA strand breaks and significantly lower levels of aneuploidy than the sperm population from which they were selected. A brand name for one such sperm selection device is PICSI. A recent clinical trial showed a sharp reduction in miscarriage with embryos derived from PICSI sperm selection.

'Washed' or 'unwashed' sperm may be used in the process.

Live birth rate are significantly higher with progesterone for luteal support in ICSI cycles. Also, addition of a GNRH agonist for luteal support in ICSI cycles has been estimated to increase success rates, by a live birth rate RD of +16% (95% confidence interval +10 to +22%).

Using ultra-high magnification during sperm selection (with the technique being called *IMSI*) has no evidence of increased live birth or miscarriage rates compared to standard ICSI.

Success or Failure Factors

One of the areas in which sperm injection can be useful is vasectomy reversal. However, potential factors that may influence pregnancy rates (and live birth rates) in ICSI include level of DNA fragmentation as measured e.g. by Comet assay, advanced maternal age and semen quality.

Complications

There is some suggestion that birth defects are increased with the use of IVF in general, and ICSI specifically, though different studies show contradictory results. In a summary position paper, the Practice Committee of the American Society of Reproductive Medicine has said it considers ICSI safe and effective therapy for male factor infertility, but may carry an increased risk for the transmission of selected genetic abnormalities to offspring, either through the procedure itself or through the increased inherent risk of such abnormalities in parents undergoing the procedure.

An online news story on MSNBC.com by Marilynn Marchione of the Associated Press, released on Saturday, May 5, 2012, discussed the risk of birth defects in ICSI versus natural conception or normal IVF; the following is a directly quoted segment of that release:

"Test-tube babies have higher rates of birth defects, and doctors have long wondered: Is it because of certain fertility treatments or infertility itself? A large new study from Australia suggests both may play a role.

Compared to those conceived naturally, babies that resulted from simple IVF, or in vitro fertilization — mixing eggs and sperm in a lab dish — had no greater risk of birth defects once factors such as the mother's age and smoking were taken into account.

However, birth defects were more common if treatment included injecting a single sperm into an egg, which is done in many cases these days, especially if male infertility is involved. About 10 percent of babies born this way had birth defects versus 6 percent of those conceived naturally, the study found." ..."

There is not enough evidence to say that ICSI procedures are safe in females with hepatitis B in regard to vertical transmission to the offspring, since the puncture of the oocyte can potentially avail for vertical transmission the offspring.

Follow-up on Fetus

In addition to regular prenatal care, prenatal aneuploidy screening based on maternal age, nuchal translucency scan and biomarkers is appropriate. However, biomarkers seem to be altered for pregnancies resulting from ICSI, causing a higher false-positive rate. Correction factors have been developed and should be used when screening for Down syndrome in singleton pregnancies after ICSI, but in twin pregnancies such correction factors have not been fully elucidated. In vanishing twin pregnancies with a second gestational sac with a dead fetus, first trimester screening should be based solely on the maternal age and the nuchal translucency scan as biomarkers are significantly altered in these cases.

References

- Blyth, Eric; Landau, Ruth (2004). Third Party Assisted Conception Across Cultures: Social, Legal and Ethical Perspectives. Jessica Kingsley Publishers. ISBN 9781843100843.
- Eric R. M. Jauniaux and Botros R. M. B. Rizk (eds.). "Introduction". Pregnancy After Assisted Reproductive Technology (PDF). Cambridge University Press. ISBN 9781107006478.
- Boulet SL, Mehta A, Kissin DM, Warner L, Kawwass JF, Jamieson DJ (2015). "Trends in use of and reproductive outcomes associated with intracytoplasmic sperm injection". JAMA. 313 (3): 255–63. doi:10.1001/jama.2014.17985. PMID 25602996.
- Van Der Linden, M.; Buckingham, K.; Farquhar, C.; Kremer, J. A. M.; Metwally, M. (2012). "Luteal phase support in assisted reproduction cycles". Human Reproduction Update. 18 (5): 473. doi:10.1093/humupd/dms017.
- Cryos International – What is the expected pregnancy rate (PR) using your donor semen? Archived January 7, 2011, at the Wayback Machine.
- Oral Sex, a Knife Fight and Then Sperm Still Impregnated Girl. Account of a Girl Impregnated After Oral Sex Shows Sperms' Incredible Survivability By LAUREN COX. abcNEWS/Health Feb. 3, 2010
- Speyer BE, Pizzey AR, Ranieri M, Joshi R, Delhanty JD, Serhal P (May 2010). "Fall in implantation rates following ICSI with sperm with high DNA fragmentation". Hum Reprod. 25 (7): 1609–1618. doi:10.1093/humrep/deq116. PMID 20495207.
- Harris, I.; Missmer, S.; Hornstein, M. (2010). "Poor success of gonadotropin-induced controlled ovarian hyperstimulation and intrauterine insemination for older women". Fertility and Sterility. 94 (1): 144–148. doi:10.1016/j.fertnstert.2009.02.040. PMID 19394605.
- Robinson, Sarah (2010-06-24). "Professor". International Federation of Gynecology and Obstetrics. Retrieved 2012-12-27.

Permissions

All chapters in this book are published with permission under the Creative Commons Attribution Share Alike License or equivalent. Every chapter published in this book has been scrutinized by our experts. Their significance has been extensively debated. The topics covered herein carry significant information for a comprehensive understanding. They may even be implemented as practical applications or may be referred to as a beginning point for further studies.

We would like to thank the editorial team for lending their expertise to make the book truly unique. They have played a crucial role in the development of this book. Without their invaluable contributions this book wouldn't have been possible. They have made vital efforts to compile up to date information on the varied aspects of this subject to make this book a valuable addition to the collection of many professionals and students.

This book was conceptualized with the vision of imparting up-to-date and integrated information in this field. To ensure the same, a matchless editorial board was set up. Every individual on the board went through rigorous rounds of assessment to prove their worth. After which they invested a large part of their time researching and compiling the most relevant data for our readers.

The editorial board has been involved in producing this book since its inception. They have spent rigorous hours researching and exploring the diverse topics which have resulted in the successful publishing of this book. They have passed on their knowledge of decades through this book. To expedite this challenging task, the publisher supported the team at every step. A small team of assistant editors was also appointed to further simplify the editing procedure and attain best results for the readers.

Apart from the editorial board, the designing team has also invested a significant amount of their time in understanding the subject and creating the most relevant covers. They scrutinized every image to scout for the most suitable representation of the subject and create an appropriate cover for the book.

The publishing team has been an ardent support to the editorial, designing and production team. Their endless efforts to recruit the best for this project, has resulted in the accomplishment of this book. They are a veteran in the field of academics and their pool of knowledge is as vast as their experience in printing. Their expertise and guidance has proved useful at every step. Their uncompromising quality standards have made this book an exceptional effort. Their encouragement from time to time has been an inspiration for everyone.

The publisher and the editorial board hope that this book will prove to be a valuable piece of knowledge for students, practitioners and scholars across the globe.

Index

www.ingramcontent.com/pod-product-compliance
Lightning Source LLC
Jackson TN
JSHW052209130125
77033JS00004B/221

* 9 7 8 1 6 3 5 4 9 0 2 6 8 *